CONCEPTS OF MODERN MATHEMATICS

Ian Stewart

DOVER PUBLICATIONS, INC.
New York

Published in Canada by General Publishing Company, Ltd., 30 Lesmill Road, Don Mills, Toronto, Ontario.
Published in the United Kingdom by Constable and Company, Ltd., 3 The Lanchesters, 162–164 Fulham Palace Road, London W6 9ER.

Bibliographical Note

This Dover edition, first published in 1995, is an unabridged, slightly corrected republication of the 1981 edition of the work first published by Penguin Books, Harmondsworth, Middlesex, England, 1975. The author has written a new Preface and some notes specially for this edition.

Library of Congress Cataloging-in-Publication Data

Stewart, Ian.
　　Concepts of modern mathematics / Ian Stewart.
　　　　　p.　　cm.
　　"An unabridged, slightly corrected republication of the 1981 edition of the work first published by Penguin Books, Harmondsworth, Middlesex, England, 1975"—T.p. verso.
　　Includes bibliographical references (p.　　–　　) and index.
　　ISBN 0-486-28424-7 (paper)
　　1. Mathematics—Popular works.　I. Title.
QA93.S73　　　1995
510—dc20　　　　　　　　　　　　　　　　　　　　　　94-39733
　　　　　　　　　　　　　　　　　　　　　　　　　　　　　CIP

Manufactured in the United States of America
Dover Publications, Inc., 31 East 2nd Street, Mineola, N.Y. 11501

Contents

Preface to the Dover Edition

Concepts of Modern Mathematics began as an extramural course—nowadays we would say 'continuing education'—taught at Warwick University in 1971. Several dozen citizens of Coventry, ranging from school students to a retired automotive engineer, gathered for two hours every week to grapple with what was then called 'Modern Mathematics' in Britain and 'New Math' in the United States. What was new about this particular style of mathematics was not its content—most of it was at least a century old—but the fact that it was being taught in schools. The abstract, general style of mathematical exposition favoured at research level by Nicolas Bourbaki—a pseudonymous group of mostly French mathematicians—was being transferred into the classroom.

This remarkable educational experiment remains controversial. Many see it as an unmitigated disaster. My own view is that its proponents confused logical issues about the nature of mathematics with psychological issues about how it should best be taught. The movement had some major advantages—for example it made a serious attempt to bring school mathematics out of the dark ages—and some serious flaws, such as the inclusion of abstract concepts that had no contact with anything that might interest the average school student. There was little communication between the educationalists who favoured introducing the new syllabus and the mathematicians who knew what the material in it was actually good for, and this led to some strange decisions.

Teachers and parents certainly found the changes very confusing, whence the course. Its aim was not to teach the new school mathematics or to discuss the issues involved in its introduction. It was to explain why the underlying abstract point of view had gained currency among research mathematicians, and to examine how it opened up entirely new realms of mathematical thought. At its best, abstract mathematics is deep, powerful, and mind-expanding. At its worst, it is superficial, obscure, and mind-numbing. The course aimed to promote the best and to ignore the worst.

'New Math' is no longer a major educational issue. Its worst excesses have been excised from the curriculum, and its positive contributions have been assimilated. Today's issues are different—the role of calculators and computers; the steady seepage of Western children into law, accountancy, medicine, advertising, and the consequent damage being done to the European and American science base; the balance between practical and theoretical skills; the importance or not of calculus . . .

Despite all these changes, the central message of *Concepts of Modern Mathematics* remains just as valid as it was twenty-plus years ago. Mathematics is a far richer area of human endeavour than most people imagine. It is versatile, powerful, and enjoyable. Even its wildest flights of fancy have important applications. This was brought home to me recently by Len Reynolds, an engineer who works at the Spring Research and Manufacturers' Association in Sheffield. Len and I are currently involved in a joint project, funded by the Department of Trade and Industry, to apply modern techniques of data analysis ('chaos theory') to the quality control of springs.

Yes, springs. Bedsprings, automobile valve springs, truck suspensions, and the little springs that you find inside ballpoint pens. What's chaotic about a spring? Almost everything! In particular, the spacing of the coils, if you measure it accurately enough. Variability of the wire translates into variable spacing of the coils, and this can cause spring manufacturers all kinds of trouble. But conventional statistics doesn't capture the qualitative type of variability that causes problems. Chaos theory, designed to extract patterns from apparently random sequences of data, does.

Importing new mathematical techniques into industry isn't straightforward. The concepts needed were not in the syllabus of the typical engineer a decade or two ago; indeed most of them are not in the syllabus now. Len asked me to recommend a simple, broad introduction to such items as set-theoretic notation, functions, and multidimensional geometry. I told him '*Concepts of Modern Mathematics*—except it's out of print,' and offered to send him a photocopy. He managed to locate a secondhand copy, telling me in passing that his supplier had reported a substantial demand for the book. Well, to be more accurate, they'd said that as soon as a copy came in, it promptly sold. I realised that maybe it would be worth getting the book back into print again.

The next step was then obvious. I own a vast collection of Dover reprints of classics that range from Arthur Cayley's *Elliptic Functions* to Jacques Hadamard's *The Psychology of Invention in the Mathematical Field,* and I asked Dover to consider giving my own humble brainchild the same treatment. I am delighted that Dover Publications has made it possible for the first book that I ever wrote for a non-specialist audience again to see the light of day. And I am convinced that its message remains as fresh as it was two decades ago.

Ian Stewart

Coventry, March 1994

Preface to the First Edition

Once upon a time it was possible for parents to help children with their homework. The 'modernization' of school mathematics has made this less possible: at the very least the parent has to learn a lot of new material, most of which seems strange and uncomfortable. A teacher friend of mine reports that his class has been clamouring to be taught 'real mathematics like Mum and Dad used to do', which sheds an interesting sidelight on where children get their opinions. Many teachers, too, find the new style of mathematics difficult to grasp hold of.

This is a pity. The aim of 'modern maths' was to encourage *understanding* of mathematics instead of blind manipulation of symbols. The true mathematician is not a juggler of numbers, but a juggler of *concepts*.

This book attempts to combat these feelings of unease. One is always uncomfortable when faced with the unknown, and the best way to lose one's fears is to see how it works, what it does, and why it does it, so that one becomes accustomed to its nature and no longer feels uncomfortable. This will not be a 'handbook of modern mathematics', but a description of the aims, methods, problems, and applications of modern mathematics: the day-to-day toolkit of the working mathematician.

I would prefer not to have assumed that the reader knows any mathematics, but here I have had to compromise. He will need a smattering of algebra, geometry and trigonometry; and the idea of a graph. I have tried to avoid calculus; it does appear occasionally but is never essential to the exposition.

Most important of all is a mind receptive to new ideas and a genuine desire to understand. Mathematics is not an easy subject – no worthwhile subject is easy – but it is a rewarding one. It is a part of our culture, and no person can count himself truly educated without some idea of what it is and does. It is, above all, a human subject, with its own triumphs and disasters, frustrations and insights.

Go on. Try it.

Acknowledgements

The quotation in Chapter 2 from *Winnie-the-Pooh* by A. A. Milne, and that at the head of Chapter 9 from *The House at Pooh Corner* by A. A. Milne are reproduced with permission of the publishers, Methuen and Co. Ltd, and Mr C. R. Milne who holds the copyright. The quotation at the head of Chapter 4 from *A Mathematician's Miscellany* by J. E. Littlewood is reproduced with permission of the publishers, Methuen and Co. Ltd. The quotation at the head of Chapter 8 from *Factor T* by Stefan Themerson is reproduced with permission of the publishers, Gaberbocchus Press Ltd. The classification of surfaces in Chapter 12 is an abridged version of that given in *Introduction to Topology* by E. C. Zeeman (to appear) and is included with Professor Zeeman's permission: however, any infelicities arising from the abridgement are my responsibility. To all of the above I express sincere thanks.

Chapter 1   Mathematics in General

> *'It is difficult to give an idea of the vast extent of modern mathematics'*
> – A. Cayley, in an address of 1883.

From the sudden conversion of our schools to 'modern mathematics' one might gain the impression that mathematics has lost control of its senses, thrown out all of its traditional ideas, and replaced them by weird and whimsical creations of no possible use to anyone.

This is not entirely an accurate picture. At a conservative estimate, most of the 'modern mathematics' now taught in schools has been in existence for over a century. In mathematics new ideas have developed naturally out of older ones, and have been incorporated steadily with the passing of time. But in our schools we have introduced a number of new concepts all at once, mostly without any discussion of how they relate to traditional mathematics.

Abstractness and Generality

One of the more noticeable aspects of modern mathematics is a tendency to become increasingly abstract. Each major concept embraces not one but many diverse objects, all having some common property. An abstract theory develops the consequences of this property, which may then be applied to *any* of the diverse objects.

Thus the concept 'group' has applications to rigid motions in space, symmetries of geometrical figures, the additive structure of whole numbers, or the deformation of curves in a topological space. The common property is the possibility of combining two objects of a certain kind to yield another. Two rigid motions, performed in succession, yield a rigid motion; the sum of two numbers is a number; two curves stuck end to end form another curve.

Abstraction and generality go hand in hand. And the main advantage of generality is that it saves work. It is pointless to prove the same theorem four times in different disguises, when it could have been proved once in a general setting.

A second feature of modern mathematics is its reliance on the language of set theory. This is usually no more than common sense in symbolic dress. Mathematics, particularly when it becomes more general, is less interested in specific objects than in whole collections of objects. That $5 = 1+4$ is not terribly significant. That every prime number of the form '$4n+1$' is a sum of two squares *is* significant. The latter is an observation about the collection of *all* prime numbers, rather than about any particular prime number.

A *set* is just a collection: we use a different word to avoid certain psychological overtones associated with the word 'collection'.[1] Sets can be combined in various ways to give other sets, in the same way that numbers can be combined (by addition, subtraction, multiplication, . . .) to give other numbers. The general theory of arithmetical operations is *algebra*: so we also can develop an algebra of set theory.

Sets have certain advantages over numbers, particularly from the point of view of teaching. They are more concrete than numbers. You cannot show a child a number ('I am holding in my hand the number 3'), you can show him a number of things: 3 lollipops, 3 ping-pong balls. You will be showing him a *set* of lollipops, or ping-pong balls. Although the sets of interest in mathematics are not concrete – they tend to be sets of numbers, or functions – the basic operations of set theory can be demonstrated by means of concrete material.

Set theory is more fundamental to mathematics than arithmetic – although the fundamentals are not always the best starting point – and the ideas of set theory are indispensable for an understanding of modern mathematics. For this reason I have discussed sets in Chapters 4 and 5. The language of set theory is used freely thereafter, though I have tried not to use anything beyond very elementary parts of the theory. It would be wrong to overemphasize set theory *per se*: it is a language, not an end in itself. If you knew set theory up to the hilt, and no other mathematics, you would be of no use to anybody. If you knew a lot of mathematics, but no set theory, you might achieve a great deal. But if you knew just *some* set theory, you would have a far better understanding of the language of mathematics.

Intuition and Formalism

The trend to greater generality has been accompanied by an increased standard of logical rigour. Euclid is now criticized because he didn't have an axiom to say that a line passing through a point inside a triangle must cut the triangle somewhere. Euler's definition of a function as 'a curve drawn by freely leading the hand' will not allow the games that mathematicians wish to play with functions, and anyway it's far too vague. (What is a 'curve'?) One can go overboard for this sort of thing, verbal arguments can be replaced by a profusion of symbolic logic and checked for validity by a blind application of a standard technique. If carried too far (and in this case, enough is too much) this destroys understanding, instead of aiding it.

The demands for greater rigour are not just a whim. The more complicated and extensive a subject becomes, the more important it is to adopt a critical attitude. A sociologist, trying to make sense of masses of experimental data, will have to discard those experiments which have been badly performed, or whose conclusions are dubious. In mathematics it is the same. All too often the 'obvious' has turned out to be false. There exist geometrical figures which do not have an area. According to Banach and Tarski[2] it is possible to cut a spherical ball into six pieces and reassemble the pieces to form two balls, *each the same size as the original.* On grounds of volume, this is impossible. But the pieces do not have volumes.

Logical rigour provides a restraining influence which is of great value in dangerous circumstances, or when dealing with subtleties. There are theorems which most practising mathematicians are convinced must be true; but until someone proves them they are unjustified assumptions, and cannot be used except as assumptions.

Another place where one must be careful about one's logic is when proving something impossible. What is impossible by one method may easily be performed by another, so very careful specifications are required. There exist proofs that quintic equations cannot be solved by radicals,[3] or that angles cannot be trisected with ruler and compasses. These are important theorems, because they close off unproductive paths. But if we are to be certain that the paths really are unproductive, we must be very cautious with our logic.

Impossibility proofs are very characteristic of mathematics. It is virtually the only subject that can be sure of its own limitations. It has at times become so obsessed with them that people have been more interested in proving that something cannot be done than in finding out how to do it! If self-knowledge be a virtue, then mathematicians as a breed are saints.

However, logic is not all. No formula ever *suggested* anything on its own. Logic can be used to solve problems, but it cannot suggest which problems to try. No one has ever formalized *significance*. To recognize what is significant you need a certain amount of experience, plus that elusive quality: *intuition*.

I cannot define what I mean by 'intuition'. It is simply what makes mathematicians (or physicists, or engineers, or poets) tick. It gives them a 'feel' for the subject; with it they can *see* that a theorem is true, without giving a formal proof, and on the basis of their vision produce a proof that works.

Practically everybody possesses some degree of mathematical intuition. A child solving a jig-saw puzzle has it. Anyone who has succeeded in packing the family's holiday luggage in the boot of the car has it. The main object in training mathematicians should be to develop their intuition into a controllable tool.

Many pages have been expended on polemics in favour of rigour over intuition, or of intuition over rigour. Both extremes miss the point: the power of mathematics lies precisely in the combination of intuition *and* rigour. Controlled genius, inspired logic. We all know the brilliant person whose ideas never quite work, and the tidy, organized person who never achieves anything worthwhile because he is too busy getting tidy and organized. These are the extremes to avoid.

Pictures

In learning mathematics, the psychological is more important than the logical. I have seen superbly logical lectures which none of the audience understood. Intuition should take precedence; it can be backed up by formal proof later. An intuitive proof allows you to understand *why* the theorem must be true; the logic merely provides firm grounds to show that it *is* true.

In subsequent chapters, I have tried to stress the intuitive side of mathematics. Instead of giving formal proofs I have tried to sketch the underlying ideas. In a proper textbook one should, ideally, do both; few texts achieve this ideal.

Some mathematicians, perhaps 10 per cent, think in formulae. Their intuition deals in formulae. But the rest think in pictures; their intuition is geometrical. Pictures carry so much more information than words. For many years schoolchildren were discouraged from drawing pictures because 'they aren't rigorous'. This was a bad mistake. Pictures are not rigorous, it is true, but they are an essential aid to thought and no one should reject anything that can help him to think better.

Why?

There are plenty of reasons for doing mathematics, and anyone reading this is unlikely to demand that the existence of mathematics be justified before he proceeds one page further. Mathematics is beautiful, intellectually stimulating – even useful.

Most of the topics I intend to discuss come from pure mathematics. The aim in pure mathematics is not practical applications, but intellectual satisfaction. In this pure mathematics resembles the fine arts – few people ask that a painting should be useful. (Unlike the fine arts it has generally accepted critical standards.) But the remarkable thing is that – almost despite itself – pure mathematics *is* useful. Let me give an example.

In the 1800s mathematicians expended a lot of energy on the *wave equation*; a partial differential equation arising from the physical properties of waves in a string or in fluid. Despite the physical origin, the problem was one of pure mathematics: nobody could think of a practical use for waves. In 1864 Maxwell laid down a number of equations to describe electrical phenomena. A simple manipulation of these equations produces the wave equation. This led Maxwell to predict the existence of electrical waves. In 1888 Hertz confirmed Maxwell's predictions experimentally, detecting radio waves in the laboratory. In 1896 Marconi made the first radio transmission.

This sequence of events is typical of the way in which pure mathematics becomes useful. First, the pure mathematician playing about with the problem for the fun of it. Second, the theoretician, applying the mathematics but making no attempt to test his theory. Third, the experimental scientist, confirming the theory but not developing any use for it. Finally the practical man, who delivers the goods to the waiting world.

The same sequence of events occurs in the development of atomic power; or in matrix theory (used in engineering and economics); or in integral equations.

Observe the time-scale. From the wave equation to Marconi: 150 years. From differential geometry to the atomic bomb: 100 years. From Cayley's first use of matrices to their use by economists: 100 years. Integral equations took thirty years to get from the point where Courant and Hilbert developed them into a useful mathematical tool to the point where they became useful in quantum theory, and it was many years after that before any practical applications came out of quantum theory. Nobody could have realized at the time that their mathematics would turn out to be needed a century or more later!

Does this mean that all mathematics, however unimportant it may seem now, should be encouraged on the off chance that it will be just what the physicists need in 2075?

The wave equation, differential geometry, matrices, integral equations: all these were recognized as significant mathematics at the time they were first developed. Mathematics has a very interrelated structure, and developments in one part can often affect other parts: this leads to a certain body of mathematics being thought of as 'central', and it is in this centre that the significant problems lie. Even totally new methods prove their significance by tackling central problems. Most mathematics that has later turned out to be useful for practical applications has come from this central region.

Mathematical intuition triumphant? Or just that any mathematics not considered significant never gets developed to the point where it *could* be useful? I don't know. But it is pretty certain that mathematics considered by a consensus of mathematicians to be trivial or unimportant will *not* prove useful. The theory of generalized left pseudo-heaps does not hold the key to the future.

However, some very beautiful and significant mathematics also turns out to be useless in practice, because the real world just doesn't work that way. A certain theoretical physicist secured himself a mighty reputation on the basis of his deduction, on very general mathematical grounds, of a formula for the radius of the universe. It was a very impressive formula, liberally spattered with es, cs, hs, and a few πs and $\sqrt{\ }$s for good measure. Being a theoretician, he never bothered to work it out numerically. It was several years before anybody had enough curiosity to substitute the numbers in it and work out the answer.

Ten centimetres.

Chapter 2   Motion without Movement

'GEOMETER: *a species of caterpillar*' –
Old Dictionary

Geometry is one of humanity's most powerful thinking tools. The visual sense dominates our perceptions, and geometrical intuition is largely visual. In geometry it is often possible to *see* (quite literally) what is going on. Pythagoras' theorem becomes almost obvious, given the pictures in Figure 1.

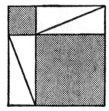 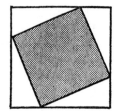

Figure 1

Furthermore, the intuitive feeling evoked by the picture can, with a little care, be turned into a logically satisfactory mathematical *proof* that the theorem is true. And because of the appeal to intuition, it is a very convincing proof.

Geometry in the style of Euclid (which until recently was the only kind of geometry that most people ever encountered) eschews pictorial arguments in favour of a stilted and essentially *algebraic* (i.e. symbol-manipulative) kind of reasoning, based on the concept of congruence of triangles and an accompanying reduction of all geometrical ideas to properties of triangles.

The notion of congruence is intuitive enough: two triangles are congruent if they have the same shape and the same size. But what children often find very difficult is the way that congruent triangles are used to prove theorems. The first 'difficult' theorem in Euclid was a notorious stumbling-block precisely because of the complicated juggling of congruent triangles in its proof. (There were other problems: in the 1850s schoolboys not only had to reproduce Euclid's proofs; they also had to use the same letters on their diagrams!)

As it happens, Euclid had several very good reasons for proceeding as he did. The overwhelming one was a wish to develop all of geometry from a few simple basic principles by strictly logical argument. It is true that later ages have found holes in the logic, but these can be filled. However, most children do not appreciate the need for logical proofs. At any stage in mathematics, one's definition of 'logically rigorous' tends to boil down to 'it convinces me'; though of course a professional logician takes a lot of convincing! A substantial part of mathematical education consists of revealing flaws in apparently convincing arguments and showing the student that he ought not to be convinced by them. If we wish to teach children geometry, we should either settle for proofs that *they* find acceptable, or we should be prepared to spend a lot of time improving their critical faculties; in which latter case a course in logic might be more helpful than a course in geometry!

But it is counter-productive to show a child a proof which is *merely* convincing, and which later turns out to be completely fallacious. The long-term effects would be confusion and distrust. We need ways of convincing the child of the truth of certain theorems which *later* can be filled out into logical proofs. The above pictures for Pythagoras's theorem are the sort of thing I mean. Before they can be made into a rigorous proof we have to work on the concept of 'area'.

In other words, the *mathematics* should reflect the *intuition*.

Euclid (whoever he was) certainly possessed a strong geometrical intuition – otherwise his book could never have been written. But he did not possess the right kind of mathematical tools to express the intuitive ideas directly, and with great ingenuity he resorted to the paraphernalia of congruence, and the rest. Mathematical developments originating in the nineteenth century have now provided such tools; the ideas involved have filtered down into the schools and are included in 'modern mathematics' programmes under the names 'transformation geometry' or 'motion geometry'.

Overturning Euclid

The theorem referred to above as 'the first difficult theorem in Euclid' is the one about isosceles triangles: *the angles at the base of an isosceles triangle are equal.* I want to begin by giving Euclid's proof of this theorem; unlike that usually given in school geometry it does not use any constructions related to the mid-point of the base. This is because when Euclid wants to prove it he does not yet have a proof that lines possess mid-points, and so cannot use the concept.

In the diagram of Figure 2 we have produced AB to a point D and AC to a point E, in such a way that $AD = AE$. We have then drawn lines DC and EB. Euclid's argument is as follows:

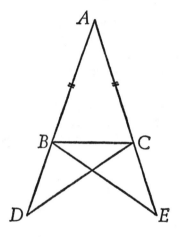

Figure 2

(i) Triangles ACD and ABE are congruent (two sides and the included angle).

(ii) Hence $\angle ABE = \angle ACD$.

(iii) Hence also $DC = EB$.

(iv) Therefore triangles DBC, ECB are congruent (three sides).

(v) Hence $\angle DCB = \angle EBC$.

(vi) From (v) and (ii) it follows by subtraction that $\angle ABC = \angle ACB$, as required.

The steps in the proof may appear more transparent if we draw a kind of strip cartoon of the main stages of the argument, as in Figure 3.

It is very striking (particularly in the cartoon) how everything comes in *pairs*. Side *AB* is on the left, *AC* on the right, and they are equal. Triangle *ACD* is on the left, *ABE* on the right, and they are congruent. And so on. Finally, ∠*ABC* is on the left, ∠*ACB* on the right: they are equal, and the theorem is proved.

This is a strong hint that if we can find a way of changing right to left and left to right, then everything should be obvious. The proof cries out for such treatment. But how can it be achieved?

Put this way, the answer is simple: *turn the triangle over*. If you make a cardboard isosceles triangle, draw round it, and then turn it over, you will find that it fits exactly. Rather than experimenting we can argue thus: if we turn it over so that *A* stays where it is and *AC* lies along the old line *AB*, then since the angle at *A* is the same measured in either direction, it follows that *AB* now lies along the old line *AC*. Since the distances *AB* and *AC* are equal, the new position of *C* is the old one of *B*, and the new position of *B* is the old one of *C*. So *B* and *C* have changed places. But now everything is determined, and all the sides fit; and new ∠*ABC* is lying on top of old ∠*ACB*, so the two are equal.

Arguments against the Motion

C. L. Dodgson, in one of his more mathematical works,[1] records the following conversation:

MINOS: It is proposed to prove [the theorem] by taking up the isosceles triangle, turning it over, and then laying it down again upon itself.

EUCLID: Surely that has too much of the Irish Bull about it, and reminds one a little too vividly of the man who walked down his own throat, to deserve a place in a strictly philosophical treatise?

MINOS: I suppose its defenders would say that it is conceived to leave a trace of itself behind, and that the reversed triangle is laid down upon the trace so left.

This disposes of one possible objection to our procedure. But there is another, deeper objection; and one which would have seemed particularly insurmountable to the ancient Greeks: the

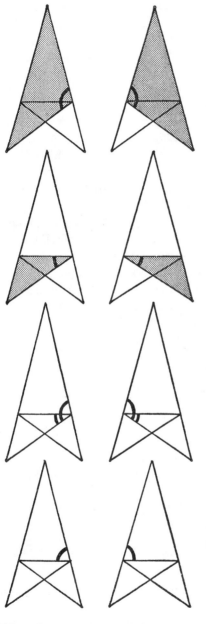

These two triangles are congruent, so the marked angles are equal.

Hence these two triangles are congruent, and the marked angles equal.

Compare the marked angles . . .

. . . and we find that these are equal.

Q.E.D.

Figure 3

whole concept of motion takes on a dubious aspect in view of Zeno's paradoxes. This may well have been the reason why Euclid turned to the safer congruence arguments.

Zeno listed four paradoxes. One will suffice here to suggest the general flavour.[2] In order to move from a point *A* to another point *B*, it is first necessary to move to a point *C* midway between. But before moving to *C*, it is necessary to move to a point *D* midway between *A* and *C*. And before moving to *D* ... It would appear that the motion can never begin!

The problem here is not as straightforward as it looks, and the ancient Greeks were well aware of the fact. In consequence, any reference to motion in a supposedly logical proof would have been considered a flaw. In the real world, of course, things *do* move; but an appeal to experimental evidence does not constitute a proof.

An Amendment to the Motion

In fact we shall sidestep completely the problems raised by Zeno's paradoxes, by a careful reformulation of our ideas.

Take hold of your cardboard triangle, turn it over, and put it back where it came from. Is it relevant to the proof on page 11 where the triangle goes in between? Does it make any difference if you flip it over deftly, or wave it around, or dance around the room to the *Blue Danube* waltz? Or if you walk out of the house, catch a train to Liverpool, hitch-hike home, and *then* put it back down?

As long as it gets put back in the same place as before, it makes no odds where it has gone in between. Indeed, it need not have gone anywhere: wave the magic wand and it just flips from one position to the other. More precisely, since it makes no difference where it goes in between, we do not need to talk about where it goes in between, and as a result we need not assume that it goes anywhere. What we *do* need to know is where each point of the triangle finishes up.

To do this we must have a way of labelling the points of the triangle, and the easiest way is to label all the points of the plane once and for all, so that we do not need to do everything all over again for a new diagram. It matters little in principle which labelling we adopt, but a particularly convenient one is furnished by coordinate geometry: each point in the Euclidean plane is labelled by its coordinates (x, y) with respect to some fixed choice of axes.

Suppose for definiteness that our axes are marked off in centimetres. Suppose we wish to move 5 cm to the right. Where does a given point (x, y) end up?

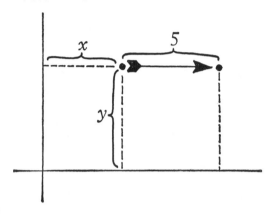

Figure 4

We can work this out from Figure 4. The y-coordinate is clearly unchanged, while the x-coordinate increases by 5. The point 5 cm to the right of (x, y) is $(x+5, y)$.

Notice now that (x, y) does not, in fact, move at all. Look at the point $(2, 3)$, then at $(7, 3)$. Did $(2, 3)$ move? Then what is it doing still sitting at $(2, 3)$? It is essential to our labelling system that the points of the plane do *not* move. What does move is our attention. If a triangle had its vertices at $(1, 1)$, $(2, 1)$, and $(1, 4)$, and *if* it moved 5 cm to the right, then its vertices would be at $(6, 1)$, $(7, 1)$, and $(6, 4)$, as in Figure 5.

What we now have is not one triangle, but two; one lying 5 cm to the right of the other. By transferring our attention from one to the other we can achieve, *without* any actual movement, the effects which would be obtained *with* movement. (Incidentally, this helps to explain Minos' idea that the triangle should leave a 'trace': we do rather more, and leave the whole triangle!)

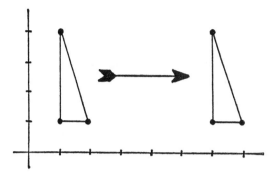

Figure 5

The way in which our attention changes can be specified by the scheme

$$(1, 1) \rightarrow (6, 1)$$
$$(2, 1) \rightarrow (7, 1)$$
$$(1, 4) \rightarrow (6, 4)$$

and in general

$$(x, y) \rightarrow (x+5, y).$$

We introduce a symbol, say T, which will mean 'the point 5 cm to the right of'. So

$$T(1, 1) = (6, 1)$$

reads: 'The point 5 cm to the right of (1, 1) is (6, 1)', and in general

$$T(x, y) = (x+5, y) \qquad (\dagger)$$

reads: 'The point 5 cm to the right of (x, y) is $(x+5, y)$'.

This new symbol T accomplishes much the same objects as the instruction 'move 5 cm to the right'. But it doesn't actually move anything; it just tells us where things would go if they did move. Furthermore, everything we need to know about T is captured in the formula ($\dagger$), which can be taken as a *definition* of T; indeed as a definition of 'move 5 cm to the right'.

Anything like T is called a *transformation* of the plane. A transformation F is considered to be known if, for every point (x, y), we know which point is $F(x, y)$. We might specify it by a formula, such as ($\dagger$); but any definite way of finding $F(x, y)$ would do. To each motion (in the intuitive sense) corresponds a transformation

F such that

$$F(x, y) = \text{the point to which an object}$$
$$\text{placed on } (x, y) \text{ would move.}$$

The transformations have the advantage that, although *motivated* by the idea of movement, they do not explicitly involve that idea, and hence avoid the taint of Zeno's paradoxes. Using transformations we can create the kind of mathematics in which the idea 'turn the triangle over and lay it down on top of itself' has a sensible interpretation, with no logical pitfalls.

Rigidity

It is instructive to work out which transformations correspond to given motions. For example, the motion 'reflect in the *x*-axis' corresponds to the transformation *G* such that

$$G(x, y) = (x, -y)$$

and 'rotate through 90° clockwise' corresponds to *H*, where

$$H(x, y) = (y, -x).$$

These can easily be read off from Figures 6 and 7.

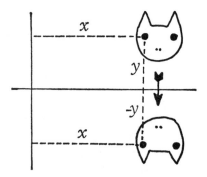

Figure 6

Conversely, we can work out which motion corresponds to a given transformation. Thus if *K* satisfies

$$K(x, y) = (x+3, y-2)$$

then the corresponding motion takes everything 3 cm to the right and 2 cm downwards.

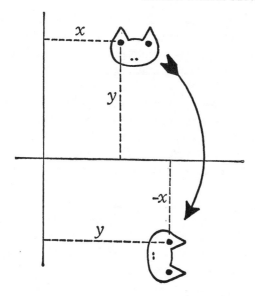

Figure 7

More complicated transformations can be investigated by plotting where a few points go. Thus if

$$J(x, y) = (x^2, xy)$$

we can compute: $J(1, 1) = (1, 1)$, $J(2, 3) = (4, 6)$, etc. and plot the resulting points on graph paper. This particular transformation takes the square with vertices $(1, 1)$, $(1, 3)$, $(3, 1)$ and $(3, 3)$ into the shape shown in Figure 8.

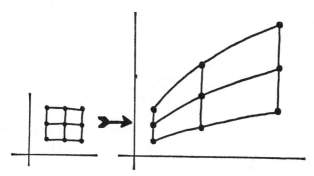

Figure 8

Thus this transformation J distorts the shape, bending and twisting it. This is not the sort of transformation that one usually wants in geometry; if we are allowed to stretch and bend things then all triangles will be interchangeable and nothing of interest results.

The sort of transformation we need for coordinate geometry corresponds to *rigid* motions, that is, movements which do not change shapes or sizes. Our argument about the isosceles triangle will not hold if the triangle changes shape when turned over. The transformations G, H, K above correspond to rigid motions, but J does not.

The essence of a rigid motion is that it does not stretch or shrink anything. No two points move closer together or further apart. In other words, points always stay the same distance from each other. We can express this algebraically by taking coordinates in the Euclidean plane. Now in coordinate geometry there is a formula for the distance between two points (x, y) and (u, v), based on Pythagoras's theorem; namely:

$$\sqrt{[(x-u)^2 + (y-v)^2]}.$$

If a transformation F is such that

$$F(x, y) = (x', y')$$
$$F(u, v) = (u', v')$$

then the distance between $F(x, y)$ and $F(u, v)$ is

$$\sqrt{[(x'-u')^2 + (y'-v')^2]}.$$

So F corresponds to a rigid motion provided the two distances are always equal, no matter which points (x, y) and (u, v) we choose. Squaring up, this means that

$$(x-u)^2 + (y-v)^2 = (x'-u')^2 + (y'-v')^2$$

for all (x, y) and (u, v). A transformation F will correspond to a rigid motion if and only if it satisfies this equation. If we wish, we may *define* a rigid motion to be such an F, thereby identifying the formal concept with the intuitive one.

In fact, by playing about with this equation, one may produce somewhat simpler characterizations of rigid motions. However, it would take us out of our way to pursue this tack. The point I want to make is that it is possible to specify a rigid motion as a special kind of transformation.

Translation, Rotation, Reflection

We now look at three special kinds of rigid motion. The *translation* (or slide) moves every point a fixed distance in a fixed direction (Figure 9).

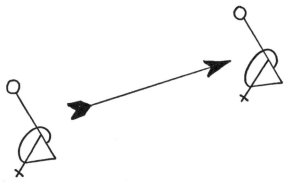

Figure 9

The *rotation*: fix a point *P* (the *centre* of rotation), and move every point around *P* through a fixed angle θ, as shown in Figure 10.

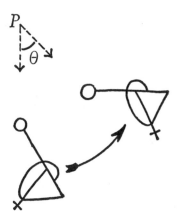

Figure 10

The *reflection*: choose a line *l*, and reflect the points of the plane as if in a mirror placed along the line (Figure 11).

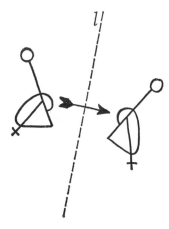

Figure 11

Using coordinate geometry we can easily work out the corresponding transformations. For example, a rotation through angle θ about the origin gives the transformation R, where

$$R(x, y) = (x \cos \theta - y \sin \theta, \ x \sin \theta + y \cos \theta).$$

From the formulae for these transformations we can also check various properties suggested by intuition: that they do indeed give rigid motions, that a rotation through θ followed by a rotation through ϕ gives a rotation through $\theta + \phi$, and so on.

We single out these three types of motion because they seem to involve quite distinct principles. Each takes a fairly simple form, as do the expressions for the corresponding transformations. We do not single out any other kinds of rigid motion, because every rigid motion of the plane can be obtained by a succession of translations, reflections, and rotations. Pictorially this follows from Figure 12.

(i) We start with a triangle ABC and any rigid motion U.

(ii) By a translation T we can make $T(A)$ and $U(A)$ coincide.

(iii) Then a rotation S about the point $U(A)$ brings $T(B)$ into coincidence with $U(B)$.

(iv) Finally reflection in the line $U(A)U(B)$ brings $S(T(C))$ into coincidence with $U(C)$.

(i)

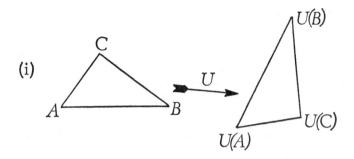

(ii)

(iii)

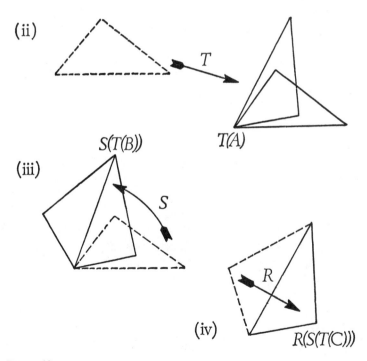

(iv)

Figure 12

(Of course, in certain cases we may not need one of these steps.)

We have used a triangle here. It is a consequence of the two-dimensionality of the plane that any rigid motion is uniquely specified by what it does to a (non-degenerate) triangle, whence it is sufficient to consider only triangles.[3] We have proved that: any rigid motion of the plane can be obtained by a translation, followed by a rotation, followed by a reflection (possibly with some stages omitted).

Obviously the reflection is needed only when the motion turns figures over. So a motion which leaves everything the same way up can be obtained as a translation followed by a rotation. (With further analysis, one can say more than this.) Suppose we apply two reflections, in (possibly) different lines. The first turns things over, the second turns them back the right way; so both reflections one after the other leave things the right way up. This means that the result of a double reflection is the same as some translation followed by some rotation. This fact is a good deal less obvious than several we have mentioned, but it emerges without difficulty from our work.

The result above can be expressed as follows: if U is any rigid motion, then there exists a translation T, a rotation S, and a reflection R, such that for any point $X = (x, y)$

$$U(X) = R\big(S(T(X))\big).$$

(We need the usual proviso that any of R, S, T may be omitted in certain cases.) A more concise notation suggests itself. Given two transformations E, F we can define EF by

$$EF(x, y) = E(F(x, y)).$$

If E corresponds to a rigid motion (which we also call E) and F to a rigid motion (F) then EF corresponds to the motion 'first do F, then do E'. (This is because when we work out $E(F(x, y))$ we first find $F(x, y)$, and *then* E of that. It is a pity that the motions come out in the wrong order;[4] a similar phenomenon occurs in evaluating log sin (x) when we first work out sin (x), then take the logarithm.)

Earlier on we had two rigid motions giving rise to transformations G, H, such that

$$G(x, y) = (x, -y)$$
$$H(x, y) = (y, -x)$$

We compute *GH* as follows:

$$GH(x, y) = G\big(H(x, y)\big)$$
$$= G(y, -x)$$
$$= (y, x).$$

(Notice that for the last line we have to remember that the symbols *x* and *y* are quite arbitrary: we could just as easily have specified that $G(u, v) = (u, -v)$, and then we can substitute *y* for *u* and *x* for *v*).

If we draw a diagram (Figure 13) we can see that this corresponds to the rigid motion obtained by reflection in the diagonal line $y = x$.

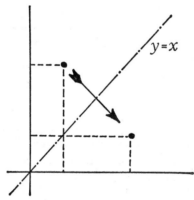

Figure 13

Call the corresponding transformation *D*, so that

$$D(x, y) = (y, x).$$

Then we have shown that

$$GH(x, y) = D(x, y)$$

for *any* point (x, y). It seems only reasonable to take this as meaning

$$GH = D.$$

If you go back to the corresponding motions and experiment, you will find that this equation checks. If you rotate through 90° clockwise and then reflect in the *x*-axis, the result is the same as a reflection in the line $y = x$. So our computation agrees with experiment.

If instead of *GH* we work out *HG*, we have

$$HG(x, y) = H\big(G(x, y)\big)$$
$$= H(x, -y)$$
$$= (-y, -x)$$

which now corresponds to a reflection in the other diagonal $y = -x$. Notice that $GH \neq HG$. There is really no reason, apart from habit, to suppose that *GH* and *HG* *should* be equal; but anyway the example shows that they need not be. This is why we must be careful as to whether *EF* means 'first *F*, then *E*' or the other way around.

We can now rewrite our equation in the very simple form

$$U = RST.$$

The fact that we can define a 'product' *EF* suggests the possibility of an 'algebra' of transformations. Pursued in one direction this idea leads to linear algebra, which we discuss in Chapter 15. In another direction it leads to group theory (Chapter 7).

Back to the Theorem

We have disgressed somewhat from the isosceles triangle which started us on our way, but we have now built up the machinery we need to make the 'turn it over' proof respectable. To someone who has had enough practice with manipulating transformations, the instruction alone will suffice. More cautiously, we should argue something like this:

There is a transformation *T* corresponding to a reflection in the line bisecting $\angle BAC$. Because rigid motions keep distances unchanged (and therefore also keep angles unchanged) it follows that $T(A) = A$, $T(B) = C$, and $T(C) = B$. Thus applying *T* to $\angle ABC$ yields $\angle ACB$. Since the sizes of angles are unchanged, we have

$$\angle ABC = \angle T(A)T(B)T(C) = \angle ACB$$

which is what we had to prove.

Once you get used to it, this really is easier to follow than Euclid's proof; and the mathematics follows the same line of argument as the intuitive idea 'turn it over'.

Now we can use the transformation concept to talk about rigid motions, without having Zeno's ghost breathing down our necks.

This opens the way to new and simplified proofs of many standard theorems of geometry. We give two examples:

(1) *If two angles of a triangle are equal, then the triangle is isosceles.*

Let the triangle be *ABC*, with angles at *A* and *B* equal. Reflect about a line perpendicular to the mid-point of *AB*. At first sight we expect a picture like Figure 14.

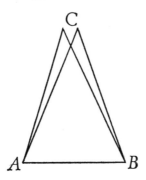

Figure 14

But equality of angles at *A* and *B* implies that the reflected triangle lies exactly on top of the unreflected one, so that *AC* is equal to *BC*. Therefore the triangle is isosceles.

(2) *Equal arcs of a circle give rise to equal chords.*

Let *A*, *B*, *X*, *Y* be points on a circle, centre *O*, with arc *AB* equal to arc *XY* (in length) as shown in Figure 15.

Rotate the figure about *O* so that *A* falls on *X*. Then because of

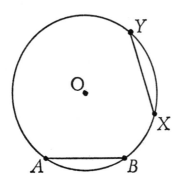

Figure 15

the equality of arc lengths B falls on Y, so chord AB falls on chord XY, and the two are equal.

You should now be able to think up other geometrical theorems which can be proved by our methods. Naturally, to carry out such a programme for the whole of geometry requires more work in setting up the basic notions, and not every geometrical theorem will be a *direct* consequence of properties of rigid motions. In fact, those which are direct consequences become essentially trivial, and we can concentrate on the less straightforward geometrical properties. Use of rigid motions helps us to sift out the really interesting results from among masses of trivia.

Chapter 3   Short Cuts in the Higher Arithmetic

> *'One of the endearing things about mathematicians is the extent to which they will go to avoid doing any real work'* – Matthew Pordage

Primitive man's consciousness of numbers may well have arisen from a desire to keep track of the important things in his life. How many sheep/arrowheads/wives do I have? How long before the spring floods? These questions focus attention on the counting numbers 1, 2, 3 . . . although the abstract concept 'number' came much later than the practical use of the idea. That two sheep and two wives have something in common – namely 'two-ness' – is by no means obvious, and is not appreciated by very young children who nevertheless can distinguish between one sheep and two sheep.

To these counting numbers other societies added other numbers, each according to its needs. The Hindus invented zero. Fractions were introduced to handle division of materials into parts. Negative numbers made their appearance; giving first the system of *integers* . . . , -3, -2, -1, 0, 1, 2, 3, . . . ; and the negative fractions, leading to the *rational numbers* p/q for integers p, q — for example, 1/2, 17/25, $-11/292$. Greek geometry and the needs of the calculus led to the *real numbers* – including numbers like $\sqrt{2}$ which cannot be represented as rational numbers – and attempts to solve algebraic equations gave rise to the mysterious *complex numbers* by insisting that -1 should have a square root and assuming that it did.

At each stage in this development there were vast intellectual battles as to whether these newfangled things really *were* numbers.

As it happens, all these numbers fit together into a grand scheme (Figure 16).

The arrows here mean that the number system at the head of the arrow contains all the numbers of the system at the tail, together with some extras.

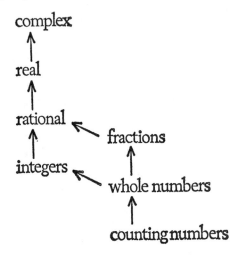

Figure 16

Further, in each system it is possible to perform 'arithmetic'. These similarities help to explain why the word 'number' was persistently attached to the objects of each of these multifarious systems. The essential arbitrariness of the entire development became forgotten, and 'number' acquired a quality akin to divine revelation.

None of the numbers in any of these systems have any existence in the real world. I have yet to meet the *number* 2 in my travels. I *have* come across 2 sheep, and insofar as I could establish at the time their behaviour was consistent with the numerical properties of that number, but I have never met the number itself. Certain properties of the real world can be described using numbers; the numbers are abstract constructs derived from real-world behaviour.

Different physical situations need different mathematical descriptions. To count how many wives we have we need only counting numbers; to weigh our gold we need fractions. A Greek geometer wanting to know the length of the hypotenuse of an isosceles right-angled triangle needed numbers like $\sqrt{2}$. A Renaissance mathematician solving a cubic[1] equation found a use for $\sqrt{-1}$.

There are many important mathematical systems which are not *called* 'numbers', as a result of historical accident and human psychology; but which arise in circumstances just as practical as those systems which *are* called 'numbers'. They often have properties in common with 'numbers' and can even be used to investigate them. The distinction between the numerical and the non-numerical is as arbitrary as the belief that 'numbers' are god-given is illusory.

Arithmetic in Miniature

A particularly interesting mathematical system is that which is sometimes referred to as *modular arithmetic*.[2] Such a system arises in any situation in which events repeat themselves in cyclic fashion: the hours of the day, the days of the week; or the measurement of angles, where 360° is the same as 0°, 361° the same as 1°, and so on.

Suppose we number the days of the week from 0 to 6, starting at Sunday, as in Figure 17.

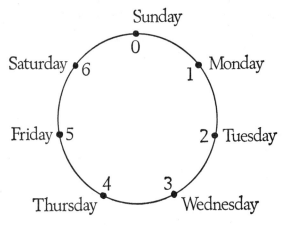

Figure 17

If we continue the numbering, day 7 is Sunday again, day 8 Monday, day 9 Tuesday .... In a certain sense we might say that

$7 = 0$, $8 = 1$, $9 = 2$, ... , where of course ' $=$ ' does not have quite its usual meaning! We can also work backwards: day -1 is the day before Sunday, which is Saturday, so $-1 = 6$; similarly $-2 = 5$. The entire system of integers becomes wrapped around the circle of days, roughly as in Figure 18.

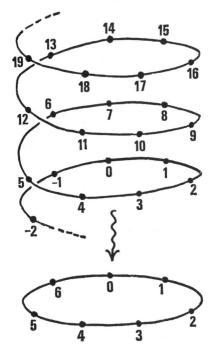

Figure 18

It is not hard to give a general criterion for which numbers fall on which day of the week:

Sunday: ..., $-14, -7, 0, 7, 14, ...$
                                   i.e. numbers of the form $7n$
Monday: ..., $-13, -6, 1, 8, 15, ...$
                                   i.e. numbers of the form $7n+1$
Tuesday: ..., $-12, -5, 2, 9, 16, ...$
                                   i.e. numbers of the form $7n+2$
Wednesday: ..., $-11, -4, 3, 10, 17, ...$
                                   i.e. numbers of the form $7n+3$

Thursday: ..., $-10, -3, 4, 11, 18, \ldots$

i.e. numbers of the form $7n+4$

Friday: ..., $-9, -2, 5, 12, 19, \ldots$

i.e. numbers of the form $7n+5$

Saturday: ..., $-8, -1, 6, 13, 20, \ldots$.

i.e. numbers of the form $7n+6$

(Numbers of the form $7n+7$, of course, are equal to $7(n+1)$ and so of the form $7n$.)

The day corresponding to a given number is determined by the remainder that number leaves on division by 7. These remainders are always 0, 1, 2, 3, 4, 5, or 6. if we indulge in a kind of 'arithmetic of remainders'. Let us agree that a statement like

$$4+5 = 2$$

carries the interpretation 'day 4 plus 5 days is day 2', which is quite a natural one, we can build up an addition table for the 'numbers' 0–6 as follows:

+	0	1	2	3	4	5	6
0	0	1	2	3	4	5	6
1	1	2	3	4	5	6	0
2	2	3	4	5	6	0	1
3	3	4	5	6	0	1	2
4	4	5	6	0	1	2	3
5	5	6	0	1	2	3	4
6	6	0	1	2	3	4	5

This table embodies the essential structure of the 7-day cycle. If we are asked 'What is 751 days after Thursday?' we rephrase it as

$$4+751 = ?$$

Now 751 isn't in our table, but we observe that

$$751 = 7.107+2$$

which is of the form $7n+2$, so that $751 = 2$. So now we have

$$4+2 = ?$$

and from the table, $? = 6$, which is a Saturday.

This 'addition' has its individual quirks: for example

$$1+1+1+1+1+1+1 = 0,$$

but interpreted in days this makes perfect sense, and one soon gets used to such peculiarities.

Emboldened by our success we might try to define *multiplication* for this system. Certainly it makes little sense to multiply Sunday by Monday; we avoid this problem by ignoring it.[3] If 3×6 is going to make any sense at all, it should be equal to $6+6+6$. From the table, this is 4. So we define

$$3 \times 6 = 4.$$

However, it would be equally reasonable to demand that 3×6 ought to be $3+3+3+3+3+3$. Perhaps this gives a different answer? But when we work it out, we get 4 again. We might argue that since $3 = 10$ then 3×6 ought to be 10×6 or 60 – but $60 = 4$. Whatever it is that we are doing, it does at least give consistent results, which is encouraging.

By use of the same repeated addition we build up this multiplication table (try it!):

×	0	1	2	3	4	5	6
0	0	0	0	0	0	0	0
1	0	1	2	3	4	5	6
2	0	2	4	6	1	3	5
3	0	3	6	2	5	1	4
4	0	4	1	5	2	6	3
5	0	5	3	1	6	4	2
6	0	6	5	4	3	2	1

The end result of our efforts – the numbers 0–6 and the two tables – is known as the system of *integers to the modulus* 7 (or *integers modulo* 7 for reasons of linguistic purity, or *integers mod* 7 for convenience). The fancy word 'modulus' is there only to signal the role played by the number 7. There is nothing special here about 7, any other number would do. If we had started with the hours on the face of a clock we would have had numbers 0–11 and arithmetic modulo 12 (or 0–23 and mod 24 for a 24-hour clock); in general any integer can serve as modulus. All you do is imagine a 'week' with that number of days, and proceed as before.

Congruences

In 1801 C. F. Gauss, reckoned to be one of the three greatest mathematicians who have ever lived, published the *Disquisitiones Arithmeticae*. This was a treatise on the theory of numbers; that is, properties of the ordinary system of integers. Gauss, of course, was interested in deeper ideas than the simple numerical computations which comprise elementary arithmetic. Number theory, from its subject-matter, may sound very easy; on the contrary it is one of the most difficult branches of mathematics and bristles with unsolved problems.

His opening section, on which all of his subsequent work was based, begins with this definition:

If a number a divides the difference of the numbers b and c, then b and c are said to be *congruent*[4] relative to a ... The number a is called the *modulus*.

(By 'number' Gauss meant 'integer'.)

If b and c are congruent to the modulus a we shall write

$$b \equiv c \qquad (\text{mod } a).$$

If it is clear from the context which modulus is being used, reference to it may be suppressed.

To see how this links up with our previous work, let's look at congruences mod 7. If b and c are congruent mod 7 then there exists an integer k such that

$$b - c = 7k$$

or that

$$b = 7k + c.$$

Thus the numbers congruent to a given number c are precisely those of the form $7k + c$. The numbers congruent to 1 mod 7 are those of the form $7k + 1$.

Given any number b we can divide it by 7 and find the remainder r, so that

$$b = 7q + r$$

from which it follows that b is congruent to r (mod 7). Since these remainders can take only values between 0 and 6, we know that every number is congruent (mod 7) to one of 0, 1, 2, 3, 4, 5 or 6.

In Figure 18, the numbers on the spiral which lie above day 0, Sunday, are those of the form $7n$; in other words, those congruent to 0. The numbers above day 1 are those congruent to 1. In general, the numbers above day d are those congruent to d.

Now it so happens that congruences can be added or multiplied, in the same way that equalities can. More precisely, if

$$a \equiv a' \qquad (\mathrm{mod}\ m)$$

and

$$b \equiv b' \qquad (\mathrm{mod}\ m)$$

then

$$a+b \equiv a'+b' \qquad (\mathrm{mod}\ m)$$

and

$$ab \equiv a'b' \qquad (\mathrm{mod}\ m).$$

Let me pause to prove this. The proof uses only elementary algebra. From the first two congruences we know that there are integers j and k such that

$$a = mj+a'$$
$$b = mk+b'. \tag{†}$$

To show that $a+b$ and $a'+b'$ are congruent, we must show that their difference

$$(a+b)-(a'+b')$$

is divisible by m. Substituting from (†) we see that this expression is

$$m(j-k)$$

which is manifestly a multiple of m. Likewise, to prove the second assertion we must look at

$$ab-a'b'$$

which reduces to

$$m(ka+jb-jkm),$$

also a multiple of m.

This means, for example, that from the facts that $1 \equiv 8$ and $3 \equiv 10$ (mod 7) we can deduce that $1+3 = 4$ is congruent to $8+10 = 18$; and that $1 \times 3 = 3$ is congruent to $8 \times 10 = 80$. As a check, the differences 14 and 77 are both divisible by 7.

Our earlier assertion that, in the arithmetic of the days of the week, $4+5 = 2$, can now be restated more accurately as:

$$4+5 \equiv 2 \qquad (\mathrm{mod}\ 7).$$

Our addition and multiplication tables are tables of congruences, rather than equalities; and an entry such as $4 \times 5 = 6$ carries the information that if any number congruent to 4 is multiplied by one congruent to 5, then the result is always congruent to 6. The arithmetic of congruences mod 7 allows us to throw away multiples of 7 if we wish, which has obvious applications in situations where moving 7 places on gets you back to the beginning.

Using congruences mod 10 it is possible to explain why all perfect squares end in 0, 1, 4, 5, 6, or 9, but not in 2, 3, 7, or 8. All numbers are congruent mod 10 to something between 0 and 9, so all squares are congruent to the squares of $0, 1, \ldots, 9$. Calculating, these are congruent to 0, 1, 4, 9, 6, 5, 6, 9, 4, 1 respectively. Since the remainder on dividing a number by 10 is its last digit (in the base 10), these last digits are all that can occur.

A great many other arithmetical occurrences can be explained in a similar way.

Division

In arithmetic mod n we can add and multiply numbers much as in ordinary arithmetic. It is also possible to subtract. The problem of *dividing* one number by another is more interesting, because the answer depends on which modulus you use.

Suppose we wish to give a meaning to 4/3 (mod 7). As yet, the symbols have no meaning; we are free to give them any meaning we wish. But we want 4/3 to have some connection with division, which limits our choice. The most natural definition would be to let 4/3 be whichever number x satisfies the equation.
$$3x \equiv 4 \qquad (\text{mod } 7).$$
If you look at the multiplication table, you will find that there is precisely one value of x which will work, namely $x = 6$. So in arithmetic mod 7 we can define
$$4/3 = 6.$$
In the same way, if p and q are any two numbers between 0 and 6, we should want p/q to be equal to y, where
$$qy \equiv p \qquad (\text{mod } 7).$$
Now qy is the number appearing in row q and column y of the multiplication table. In order for the congruence to have a

solution y, the number p must occur somewhere in row q. And in order to have a *unique* solution, p must occur only *once* in row q. (If there are two or more solutions, we don't know which one to take for p/q.)

The multiplication table (mod 7) is such that in every row *apart from row 0* each number occurs once and only once. So we can find a unique solution of the above congruence for any *non-zero q*. This means that we can define p/q when $q \neq 0$. This is not much of a restriction, because we don't expect to be able to divide by zero in any case.

If instead we work to the modulus 6, what happens? The multiplication table now takes the form

×	0	1	2	3	4	5
0	0	0	0	0	0	0
1	0	1	2	3	4	5
2	0	2	4	0	2	4
3	0	3	0	3	0	3
4	0	4	2	0	4	2
5	0	5	4	3	2	1

with very different consequences. Only in rows 1 and 5 does each number appear. In row 2, only the numbers 0, 2, and 4 appear, each twice. In row 3, only 0 and 3 appear. So we can divide by 1 or by 5 without any difficulty. But there is no way to define 1/2, or 3/4. And there are two different candidates for 4/2 (namely 2 and 5), and *three* candidates for 3/3. What a mess! How different from the modulus 7!

There is no good way out of this dilemma. One must accept that in the modulus 6, division is not always possible; the situation is a good deal worse than what happens for the integers (not to any modulus). Although we cannot divide one integer by another and get an *integer*, in general, we can enlarge the system of integers into the rational numbers and do division in the larger system. Moreover, the larger system satisfies all the 'laws of arithmetic' (such as $a+b = b+a$) which the system of integers does.

We *cannot* enlarge the system of integers mod 6 into anything where division is possible, and the laws of arithmetic still hold.

(We shall have much more to say about these 'laws' in Chapter 6.) By 'enlarge' I mean 'add a few more "numbers" '. Note that the integers mod 6 cannot be enlarged to get the ordinary integers, because that would involve changing the multiplication table and the addition table; instead of enlarging the system we destroy it.

This is because there are too many 0s in the multiplication table. There are cases where the product of two non-zero numbers is zero: for example

$$2 \times 3 \equiv 0 \qquad (\text{mod } 6).$$

Suppose we could enlarge the system into one where we could define 1/2 to be something – say a. Then, using the laws of arithmetic, and the notation ' . ' for ' $\times$ ', we have

$$3 \equiv 1.3 \equiv (a.2).3 \equiv a.(2.3) \equiv a.0 \equiv 0 \qquad (\text{mod } 6)$$

which isn't true. So if we want an enlarged system, it won't satisfy the laws of arithmetic.

The same trouble will arise in any modulus m where the product of two non-zero numbers can be zero.

By writing out multiplication tables to the moduli 2, 3, 4, 5, . . . it becomes apparent that division (except by zero) is always possible if the modulus is 2, 3, 5, 7, 11, 13, 17, . . . and is not always possible if it is 4, 6, 8, 9, 10, 12, 14, 15, 16, . . . It doesn't take a genius to recognize the pattern. The first list of numbers looks like the sequence of *prime* numbers (with no divisors other than themselves and 1); the second seems to be the *composite* numbers (which can be expressed as a product of two smaller ones).

We can easily prove that with a composite modulus, division is not always possible. Suppose the modulus is m, where $m = a.b$ and each of a and b is smaller than m. Then neither a nor b is congruent to 0 (mod m), but their product $a.b$ is m, which *is* congruent to 0. We noticed earlier that $2.3 \equiv 0$ (mod 6), and this is just a more general case of the same thing. In the same way that we deduce that 1/2 cannot be defined, it follows that (mod m) we can't define $1/a$ (or $1/b$) in any useful way.

This disposes of the composite moduli. What of the prime moduli? For all we know, there could be some prime moduli where division is not always possible. Our evidence only covers the first few primes, but perhaps for a very big prime (maybe too big to compute a multiplication table) something different happens.

Let us take a prime number p. Let t be anything not congruent to zero (mod p). Recall that division by t will be possible provided each number (mod p) occurs exactly once in row t of the multiplication table. Let us first establish that no number occurs *twice*. If it did, there would be two distinct numbers (mod p), say u and v, such that

$$tu \equiv tv \qquad (\mathrm{mod}\, p)$$

whence

$$t(u-v) \equiv 0 \qquad (\mathrm{mod}\, p).$$

So, reverting to ordinary integers, the product $t(u-v)$ is divisible by p. But if a *prime* number divides a product of two numbers, then it must divide one of them. If p divides t, then $t \equiv 0$ (mod p), which is impossible by our choice of t. If p divides $(u-v)$ then $u \equiv v$ (mod p), which is also impossible. So our assumption that the same number occurs twice in row t leads to an impossibility. It must, therefore, be a false assumption. We are left with only one possibility: no number occurs twice in row t.

There are exactly p spaces in row t, and exactly p different numbers (namely $0 \ldots p-1$) which can appear. If we can't put any in twice, then the only way to make up the numbers is to put *each* one in *once*. (This is known as the 'pigeon-hole principle'.) So each number occurs exactly once in row t. By what we said before, this implies that we can define division by t in a unique manner.

Here is an amusing application, to the famous 'Fermat numbers'. In 1640 Fermat asserted[5] that all numbers of the form

$$2^{2^n}+1$$

are prime, but remarked that he could not prove this. The first few are 3, 5, 17, 257, and 65 537, which are prime. Euler, in 1732, showed that Fermat was wrong; the next number in the sequence is $2^{32}+1$, and this is divisible by 641. Euler found this by explicit calculation. But once we know what the answer is, there's an easier way.

Observe that 641 is prime, and that $641 = 2^4+5^4 = 1+5.2^7$. Working to the modulus 641 we have:

$$2^7 \equiv -1/5$$

so that

$$2^8 \equiv -2/5,$$

whence

$$2^{32} \equiv (-2/5)^4$$
$$\equiv 2^4/5^4$$
$$\equiv -1$$

(using the first equality above for 641). Therefore $2^{32}+1$ is divisible by 641.

Two Famous Theorems

Congruences can be used for more than just numerical calculations. They are particularly important in the theory of numbers; and I shall illustrate this by proving two famous theorems. The proofs are not hard to understand once you see them, but, as E. T. Bell[6] has said, '. . . It is safe to wager that out of a million human beings of normal intelligence of any or all ages, less than ten of those who had no more mathematics than grammar-grade arithmetic would succeed in finding a proof within a reasonable time – say a year.'

If you work out successive powers of numbers to the modulus 7, you will find that they repeat the same sequence over and over again. For instance, the powers of 2 are

$2^0 \equiv 1$	$2^3 \equiv 1$	$2^6 \equiv 1$
$2^1 \equiv 2$	$2^4 \equiv 2$	$2^7 \equiv 2$
$2^2 \equiv 4$	$2^5 \equiv 4$	$2^8 \equiv 4 \dots \pmod 7$

and the pattern 1, 2, 4, 1, 2, 4, 1, 2, 4 repeats for ever. For powers of 3 the pattern is 1, 3, 2, 6, 4, 5 repeated; and there are similar patterns for the other numbers, as you can easily check for yourself.

It is easy to see that once some power becomes equal to 1, the sequence must repeat. Since $3^6 \equiv 1$, it follows that $3^7 \equiv 3^1$, $3^8 \equiv 3^2$, etc. To the modulus 7, every number other than 0 satisfies

$$x^6 \equiv 1 \pmod 7$$

(although for certain values of x something less than 6 will also work).

If you do the necessary arithmetic, you will see that for the modulus 5 every non-zero number satisfies

$$x^4 \equiv 1 \pmod 5,$$

for modulus 11 the result is
$$x^{10} \equiv 1 \qquad (\text{mod } 11)$$
and for 13
$$x^{12} \equiv 1 \qquad (\text{mod } 13).$$
I'm restricting attention to prime moduli because the pattern is more obvious there. It looks very much as if we should have
$$x^{p-1} \equiv 1 \qquad (\text{mod } p)$$
for any prime p and any x not congruent to 0 (mod p).

One way to prove this can be illustrated by working modulo 7. The non-zero numbers mod 7 are
$$1\ 2\ 3\ 4\ 5\ 6.$$
If we double all these we get
$$2\ 4\ 6\ 1\ 3\ 5,$$
which are the same numbers in a different order. So the products
$$1.2.3.4.5.6$$
and
$$2.4.6.1.3.5$$
are congruent mod 7. But the second one is also congruent to
$$(1.2).(2.2)(3.2)(4.2)(5.2)(6.2) \qquad (\text{mod } 7)$$
which is
$$2^6.(1.2.3.4.5.6). \qquad (\text{mod } 7)$$
Therefore
$$1.2.3.4.5.6 \equiv 2^6.(1.2.3.4.5.6) \qquad (\text{mod } 7)$$
and dividing out we get
$$1 \equiv 2^6 \qquad (\text{mod } 7).$$
The same thing happens if we triple all the numbers: we now get
$$3\ 6\ 2\ 5\ 1\ 4$$
and the same argument leads to
$$1 \equiv 3^6 \qquad (\text{mod } 7).$$
Now we'll do the general case mod p. Since p is prime we know that every number appears exactly once in row x of the multiplication table mod p. So the numbers
$$(1.x), (2.x), \ldots, ((p-1).x)$$
are just $1, \ldots, p-1$ in a different order. Multiplying them all together gives
$$x^{p-1}(1.2\ldots(p-1)) \equiv 1.2\ldots(p-1) \qquad (\text{mod } p)$$

so dividing both sides by $1.2 \ldots (p-1)$ we get
$$x^{p-1} \equiv 1 \qquad (\bmod\ p)$$
and the theorem is proved.

A simple application of this theorem tells us that
$$7^{18} - 1 = 1\ 628\ 413\ 597\ 910\ 448$$
is divisible by 19, without our doing any division. In deeper number-theoretic work, the general theorem is indispensable.[7] It is known as *Fermat's theorem* (not to be confused with Fermat's last theorem![8]).

The second theorem tells us more about the product
$$1.2 \ldots (p-1)$$
which occurred in the proof of Fermat's theorem. Can we evaluate this mod p?

When $p = 7$ the product is
$$1.2.3.4.5.6.$$
If we rewrite this as
$$1.(2.4)(3.5).6$$
we find that it is congruent to
$$1.1.1.(-1)$$
which is -1. The pairing off of numbers is chosen to make the products of pairs 1.

We can do the same sort of thing mod 11:
$$1.2.3.4.5.6.7.8.9.10$$
$$= 1.(2.6)(3.4)(5.9)(7.8).10$$
$$= 1.1.1.1.1.(-1)$$
$$= -1.$$

Or mod 13:
$$1.2.3.4.5.6.7.8.9.10.11.12$$
$$= 1.(2.7)(3.9)(4.10)(5.8)(6.11).12$$
$$= 1.1.1.1.1.1.(-1)$$
$$= -1.$$

In the general case we take the numbers $1, 2, \ldots, p-1$ and pair off each number with its reciprocal. This cancels out all the numbers except those which equal their reciprocals: these satisfy
$$x \equiv 1/x \qquad (\bmod\ p)$$

or
$$x^2 \equiv 1 \qquad (\text{mod } p)$$
which is the same as
$$x^2 - 1 \equiv 0 \qquad (\text{mod } p)$$
which factorizes:
$$(x-1)(x+1) = 0 \qquad (\text{mod } p).$$
So either $x \equiv 1$ or $x \equiv -1$. So we can rewrite
$$1.2 \ldots (p-1) \equiv 1.(?.?) \ldots (?.?).(-1)$$
$$\equiv -1.$$
This proves that for any prime number p,
$$1.2 \ldots (p-1) \equiv -1 \qquad (\text{mod } p)$$
which is known as *Wilson's theorem*.

If instead of p we take a composite number m, the theorem is false. For if m is composite, it has some factor $d \leq m-1$. Then d will divide $1.2 \ldots (m-1)$, and so will leave remainder 1 on dividing $1.2 \ldots (m-1)+1$. This means that m cannot divide $1.2 \ldots (m-1)+1$.

Theoretically we have a test for prime numbers. To find out whether a given number q is prime, we work out
$$1.2 \ldots (q-1)+1$$
and divide by q. If there is no remainder, q is prime: if there is a remainder, q is composite. Thus
$$1.2.3.4.5.6+1 = 721$$
is divisible by 7, so 7 is prime;
$$1.2.3.4.5+1 = 121$$
is not divisible by 6, so 6 is composite.

However, even for a relatively small number like 17 we have to work out $1.2 \ldots 16+1$, which is 20 922 789 888 001, and divide by 17. The test is not a practical one, even on a fast computer.

But it is a striking theoretical result: that one could test for primes *without* trying possible divisors.

Chapter 4   The Language of Sets

*'I have had occasion to read aloud
the phase "where E′ is any dashed [i.e.
derived] set". It is necessary to place the
stress with care'* – J. E. Littlewood

Almost any book on 'modern mathematics' talks about *sets*, and
is liberally bespattered with strange symbols like $\in$, $\subseteq$, $\cup$, $\cap$, $\emptyset$.
The present volume will be no exception, although I shall try to
keep the symbols to a minimum. There is a good reason for this
obsession with sets. Set theory is a language. Without it, not only
can we not *do* modern mathematics, we can't even say what we are
talking about. It is like trying to study French literature without
knowing any French. We shall need to use some of the language
of set theory in the remainder of the book; hence this chapter.

    A *set* is a collection of objects: the set of all English counties,
the set of all epic poems, the set of all red-headed Irishmen. The
objects belonging to the set are the *elements* or *members* of the set
(we shall use both terms indiscriminately). Thus *Paradise Lost* is
a member of the set of all epic poems; Kent is an element of the
set of all English counties. Although in introducing set theory it is
helpful to work with concrete sets, whose members are real
objects, the sets of interest in mathematics always have members
which are abstract mathematical objects: the set of all circles in
the plane, the set of points on a sphere, the set of all numbers.

    Many of the concepts of set theory can be vividly illustrated
using simple apparatus: a few small objects (pencil, eraser, pencil-
sharpener, some marbles, a sugar mouse, etc.) The objects (or
some of them) will be the elements of the set; the set itself will
consist of the chosen objects *inside a bag*. (It is crucial to have the
bag.) To find out whether or not a given object is a member of the
set, you *look in the bag and see*. For this reason, polythene bags
are best! You may find it helpful to have such apparatus on hand
for what follows.

    We shall build up an *algebra* of sets. As in ordinary algebra, we
shall use letters to represent sets and elements. To help keep track
we shall generally use small letters for elements and capital letters

for sets, but it is impossible to keep rigidly to this convention because sets can themselves be elements of other sets (put one bag inside another!) If S is the set of all epic poems, and if x is *Paradise Lost*, then x is a member of S. The phrase 'is a member of' occurs so often that it is convenient to have a symbol; the one currently in use is ∈.[1] So

$$x \in S$$

means 'x is a member of S'.

A set is considered to be known if we know what its elements are – or at any rate if in theory we can find out. There are many ways of specifying a set, of which the simplest is to list all the members. In this way the electoral roll defines the set of people entitled to vote. The standard notation for this is to enclose the list in curly brackets. So {1, 2, 3, 4} is the set whose members are 1, 2, 3, 4 *and only these*, while {spring, summer, autumn, winter} is the set of seasons. Figure 19 shows the set

{pencil, marble, sugar mouse}.

The curly brackets play exactly the role of the polythene bag.

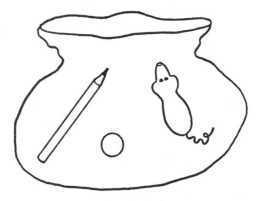

Figure 19

Two sets are *equal* if they have the same elements. Although we can put two pencils into a polythene bag, we cannot put the same pencil in *twice* (unless we take it out in between). Unhappily, there is no such physical restriction on our curly bracket notation; we can easily write things like {1, 2, 3, 4, 4, 4}. Read literally, this is the set whose members are 1, 2, 3, 4, and 4, and 4. There is a passage in *Winnie-the-Pooh*[2] where Pooh keeps saying, 'Oh, and Eeyore. I keep forgetting about him,' when Rabbit is listing the inhabitants of the forest. Despite being mentioned several times, there is only one Eeyore in the forest. In the same way, although we may list the number 4 several times, there is only one 4 in the set; which is thus equal to $\{1, 2, 3, 4\}$. When using the curly bracket notation, elements listed more than once are thought of as occurring once only in the set.

Again, there is no particular order to the objects in a bag. The curly-bracket notation introduces an artificial ordering because of the convention that we read from left right. The set $\{1, 3, 2, 4\}$ has the same elements as $\{1, 2, 3, 4\}$, so is the same set. The order inside the brackets makes no difference.

You may ask, 'But what if I want to put *two* pencils in the set?' If they are different pencils, there is no problem: put them both in. Since they are different, you are not putting anything in *twice*, just two similar objects once each. If they are the same, you haven't got two pencils.

These conventions are eminently reasonable. If your name appears twice on the electoral roll, does that entitle you to two votes? Does the order on the roll carry any electoral privileges?

More generally, a symbol such as

$$\{\text{all epic poems}\}$$

denotes the set of all epic poems. A variation of this idea allows us to write

$$\{x \mid x \text{ is an epic poem}\}$$

for the same set. The vertical bar may be read as 'such that'; and the set of all x such that x is an epic poem is the same as the set of all epic poems. The set

$$\{n \mid n \text{ is an integer and } 1 \leq n \leq 4\}$$

is the same as the set

$$\{1, 2, 3, 4\}.$$

Instead of a list, we give a property which specifies precisely the elements we wish to be included in the set. If we are careful with our definitions, making sure that we specify the *exact* property we want, this is as good as a list, and is usually more convenient. For sets with infinitely many members, such as $\{\text{all whole numbers}\}$, it is in any case impossible to give a complete list. The same is true for sets with a sufficiently large finite set of elements.

The word 'collection' has unfortunate overtones (which is why we introduce the word 'set'). The mathematical notion of a set allows sets with only one member – or even no members at all – whereas 'collections' usually have lots of members. If you were asked to look at a stamp collection containing only one stamp you might feel unimpressed. (On the other hand, if that stamp were a certain one-cent black-on-magenta British Guiana stamp of 1856, of which only one copy is known . . .) Now if we specify a set by some property it may turn out later that there is only one object with that property, or none at all. But often this is not apparent when the set is specified; and it would be stupid to have 'sets' floating around which might or might not, on looking hard enough, actually *be* sets. Thus $\{n \mid n$ is a whole number greater than 1 such that the equation $x^n + y^n = z^n$ has a solution in non-zero integers $x, y, z\}$ has at least one member, namely 2. But nobody has any idea whether or not it has any more. It is a very difficult problem in number theory[3] which has been unsolved for over 300 years. Whether or not this is a set should not have to depend on solving the problem; but it may turn out that 2 is the only member. So we must allow sets to have only one element, if that's what turns out to happen.

Sets with one element must not be confused with the element itself. It is not true that x and $\{x\}$ are equal. This is easily seen using bags (Figure 20):

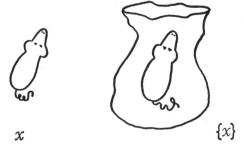

x $\{x\}$

Figure 20

and is confirmed by the observation that $\{x\}$ has just one member, namely x, while x may have any number of members depending on whether or not it is a set, and if it is, which set.

The Empty Set

For exactly the same reasons that we allow sets with just one element, we have to allow sets with no elements at all. The set of all unicorns at present residing in Bexhill is, to the best of my knowledge, such a set.

A set with no elements is called an *empty* set. (Think of an empty polythene bag.)

A fact now emerges which many people find surprising: there is only one empty set. All empty sets are equal. Perfect democracy prevails. Recall that two sets are equal if they have the same members. If they are unequal, then they do not have the same members, so one of them must have at least one member that the other does not have. In particular, one of them must have a member. If they are both empty, this is not the case; so they are not unequal. Therefore they must be equal.

This may seem odd. It is an example of 'vacuous reasoning', where the desired property holds, as it were, by default. Often trivial ideas are hard to grasp. One assumes one is looking for something substantial when in truth there is nothing there, and so one believes that one has failed to see what one was looking for. Any two empty sets are equal because, in the absence of any members to distinguish them by, there is no way to tell them apart. The contents of two empty bags are identical.

A sugar mouse has no members. Am I asserting that a sugar mouse is equal to the empty set?

Indeed I am not. The proof applies only when we have two *sets*. The best I can say is that *if* a sugar mouse is a set, and *if* (as seems likely) it has no members, *then* it is equal to the empty set.

Having established that there is just one empty set we can give it a symbol: the current one being

$$\emptyset$$

which is not a Greek phi (ϕ) but a special symbol concocted from a '0' and a '/'. The empty set is not 'nothing'; nor does it fail to exist. It is just as much in existence as any other set. It is its *members* that do not exist. It must not be confused with the number 0: for 0 is a number, whereas $\emptyset$ is a set.[4]

$\emptyset$ is one of the most useful sets in mathematics. One of its uses is to express concisely that something does not happen. Let U denote the set of unicorns in Bexhill. Then

$$U = \emptyset$$

tells us that there are no unicorns in Bexhill.

Subsets

Often one set is part of (as distinct from a member of) some other set. The set of all women is part of the set of all human beings; the set of all even numbers is part of the set of all whole numbers. Again the phrase 'part of' has unfortunate connotations, and mathematicians have been forced to invent a new word to denote the precise concept involved.

A set S is said to be a *subset* of a set T provided that every member of S is a member of T. Every member of the set W of all women is a woman, hence a human being, hence a member of the set H of all human beings. So W is a subset of H. We use the notation[5]

$$W \subseteq H$$

to describe this, so the symbol $\subseteq$ should be read as 'is a subset of'. We also say that W is *contained in H*.

A 'bag' picture for the concept 'subset' is more contrived than our earlier examples. If S is a set consisting of a pencil and an eraser, while T consists of the same pencil and eraser, plus three marbles, then the arrangement of bags in Figure 21 is decidedly misleading.

It looks as if we have just one set, whose members are

(i) three marbles,
(ii) a set whose elements are a pencil and an eraser.

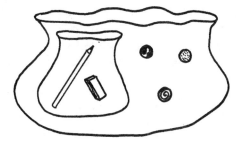

Figure 21

Figure 22 is a better picture; but to get this working practically one needs an arrangement of interpenetrating bags. (These are indicated by dotted lines.)

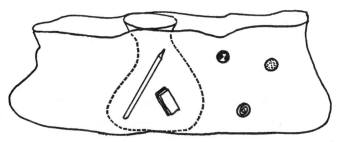

Figure 22

Certain facts follow at once from our definition of 'subset'. Every set is a subset of itself – because all of its members are members of it. Further, the empty set $\emptyset$ is a subset of any set you care to name – by another piece of vacuous reasoning. If it were not a subset of a given set S, then there would have to be some element of $\emptyset$ which was not an element of S. In particular there would have to be an element of $\emptyset$. Since $\emptyset$ has no elements this is impossible. (These are two reasons why the phrase 'part of' is misleading: the part may be the whole thing, or it may be empty.)

One nice property of subsets is that a subset of a subset is itself a subset: if $A \subseteq B$ and $B \subseteq C$ then $A \subseteq C$. For if every element of A is an element of B, and if every element of B is an element of C, then every element of A is an element of C.

Our remarks about the 'systems' of numbers at the start of Chapter 3 is really about sets and subsets. We introduce some standard notation[6] for sets of numbers (which we shall use consistently throughout). To remind you that it is standard, we use bold type.

> **N** is the set of whole numbers 0, 1, 2, 3, ...
>
> **Z** is the set of integers . . . −2, −1, 0, 1, 2, ...
>
> **Q** is the set of rational numbers (of the form p/q where p and q are integers and $q \neq 0$).
>
> **R** is the set of real numbers (representable by infinite decimals, not necessarily recurring: numbers like $\sqrt{2}$ or π).
>
> **C** is the set of complex numbers (which we won't use much, but it should be mentioned here).

These sets are among the 'systems' mentioned. The 'grand scheme' into which they all fit can be expressed as:
$$\mathbf{N} \subseteq \mathbf{Z} \subseteq \mathbf{Q} \subseteq \mathbf{R} \subseteq \mathbf{C}.$$
By the above remark it follows that $\mathbf{N} \subseteq \mathbf{Q}$, or that $\mathbf{Z} \subseteq \mathbf{R}$, and the like.

It is important not to confuse $\subseteq$ with $\in$. The two concepts have little in common. The *subsets* of $\{1, 2, 3\}$ are the sets $\emptyset$, $\{1\}$, $\{2\}$, $\{3\}$, $\{1, 2\}$, $\{1, 3\}$, $\{2, 3\}$, and $\{1, 2, 3\}$. The *elements* of $\{1, 2, 3\}$ are 1, 2, and 3. Further, it is *not* the case that if $A \in B$ and $B \in C$ then $A \in C$.[7]

Unions and Intersections

Sets may be combined together to give other sets. Out of a conceivably infinite number of possible ways of combining them, a very small number have been found useful. Prominent among these are the union and intersection of sets.

The *union* of two sets S and T is the set whose elements are those of S, together with those of T. We use the symbols
$$S \cup T.$$
Thus if $S = \{1, 3, 2, 9\}$ and $T = \{1, 7, 5, 2\}$ then
$$S \cup T = \{1, 3, 2, 9, 7, 5\}.$$

If

$$P = \{\text{all women under 35 years of age}\}$$
$$Q = \{\text{all bus-conductors}\}$$

then $P \cup Q$ will be the set of all people who are either women under 35 years of age, or bus-conductors (including those who are both).

In a similar fashion the *intersection*

$$S \cap T$$

is the set whose members are the elements which are *common* to S and T. In the above examples we have

$$S \cap T = \{1, 2\}$$

$P \cap Q = \{\text{all women bus conductors aged under 35}\}$.

With polythene bags, and the two sets S and T of Figure 23

Figure 23

then $S \cup T$ is the set obtained by putting all the objects into a single bag (Figure 24)

Figure 24

while $S \cap T$ consists of the objects which lie in both bags simultaneously (Figure 25).

Figure 25

Instead of drawing the bags from the side, we might instead use a top view (Figure 26).

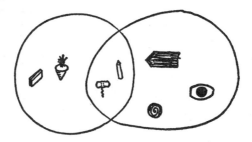

Figure 26

Then $S \cup T$ and $S \cap T$ are the sets of objects in the shaded regions of Figure 27.

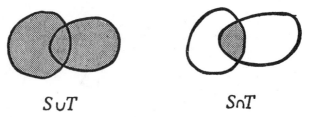

$S \cup T$ $S \cap T$

Figure 27

Now we can forget about the contents of the bags. A general picture of $S \cup T$ and $S \cap T$, for any two sets S and T, will be as shown in Figure 28.

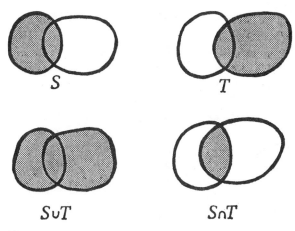

Figure 28

Diagrams like these, with circles to represent bags (i.e. sets) and shading to indicate where the relevant objects (elements) lie, are known as *Venn diagrams* after their inventor.

The symbols $\cup$ and $\cap$[8] obey various general laws, in the same way that addition and multiplication of numbers obey certain general laws. For example, whatever sets we take for A and B, it is always the case that

$$A \cup B = B \cup A$$
$$A \cap B = B \cap A.$$

For $A \cup B$ consists of all the elements of A and all the elements of B, which is the same as all the elements of B and all the elements of A. If you draw a Venn diagram, then $A \cup B$ and $B \cup A$ are both the region obtained by shading in both circles representing A and B. Similarly for $A \cap B$, only now the region will be that common to A and B.

If A, B, C are any three sets, then

$$(A \cup B) \cup C = A \cup (B \cup C)$$
$$(A \cap B) \cap C = A \cap (B \cap C).$$

The first of these says that if we combine together the elements of the three sets, it doesn't matter in which order we do it; the second that if we take the elements common to all three sets, again the order is irrelevant. You could draw the Venn diagrams: this time you will need three overlapping circles. I'll illustrate the method on yet another law.

There are two laws which connect together the operations $\cup$ and $\cap$. For any three sets A, B, C, we have

$$(A \cup B) \cap C = (A \cap C) \cup (B \cap C)$$
$$(A \cap B) \cup C = (A \cup C) \cap (B \cup C).$$

The first of these is shown by the Venn diagrams of Figure 29.

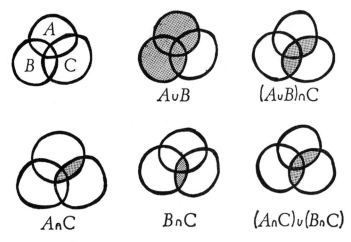

$A \cup B$ $(A \cup B) \cap C$

$A \cap C$ $B \cap C$ $(A \cap C) \cup (B \cap C)$

Figure 29

Instead of using Venn diagrams to demonstrate set-theoretic laws, we could use *membership tables*.[9] An element is in $S \cup T$ if it is in S, or in T, or both. It is in $S \cap T$ if it is in S and in T. If we write 'I' for 'in' and 'O' for 'out' these remarks can be summed up by two tables:

S	T	$S \cup T$
I	I	I
I	O	I
O	I	I
O	O	O

S	T	$S \cap T$
I	I	I
I	O	O
O	I	O
O	O	O

(For example, the third line in the $S \cup T$ table reads: 'If an element is out of S and in T, then it is in $S \cup T$'.)

To prove the second of the above laws connecting $\cup$ and $\cap$,
$$(A \cap B) \cup C = (A \cup C) \cap (B \cup C),$$
we consider the eight different possible ways elements can be in or out of A, B, and C; for each of these we tabulate whether or not it is then in, or out of, $(A \cap B) \cup C$ and $(A \cup C) \cap (B \cup C)$.

A	B	C	$A \cap B$	$(A \cap B) \cup C$
I	I	I	I	I
I	I	O	I	I
I	O	I	O	I
I	O	O	O	O
O	I	I	O	I
O	I	O	O	O
O	O	I	O	I
O	O	O	O	O

And

A	B	C	$(A \cup C)$	$(B \cup C)$	$(A \cup C) \cap (B \cup C)$
I	I	I	I	I	I
I	I	O	I	I	I
I	O	I	I	I	I
I	O	O	I	O	O
O	I	I	I	I	I
O	I	O	O	I	O
O	O	I	I	I	I
O	O	O	O	O	O

Notice that the two final columns are the same. So an element is in $(A \cap B) \cup C$ if it is in $(A \cup C) \cap (B \cup C)$, and is out of $(A \cap B) \cup C$ if it is out of $(A \cup C) \cap (B \cup C)$. But this implies that the two sets are equal, and proves the result.

With Venn diagrams (once you understand in what way they represent *general* sets) you can *see* why an identity is true. With membership tables you can *prove* it.

Complements

Another useful method of combining two sets A and B is to take their *difference*

$$A - B$$

which consists of those elements which are in A but not in B. In a Venn diagram it looks like Figure 30.

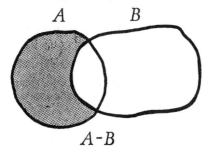

Figure 30

The corresponding membership table is

A	B	$A-B$
I	I	O
I	O	I
O	I	O
O	O	O

The *complement* S' of a set S is the set of all elements which do not belong to S. If we let V be the set of all possible elements – everything, of any kind, which could be a member of some set – then $S' = V-S$. So S' is represented by the shaded region in Figure 31.

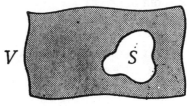

Figure 31

I find this set V somewhat daunting to contemplate. It contains so many things! All possible numbers, dogs, cats, people, books, . . . all possible concepts . . . , all sets. In fact, since V is a possible element, V is a member of V. In many respects V is *too large*. In a discussion about dogs, if one wishes to talk about all non-sheepdogs, it is pointless to worry about camels.

In any particular problem, the sets one is concerned with often lie inside some reasonably small *universal set*. If we are talking about dogs, we could take our universal set to be the set of all dogs. It might be more convenient instead to use the set of all animals. There is no fixed way of choosing a universal set. But once we have chosen one we can use it instead of V. The complement S' will consist of those elements of the universal set which are not in S; that is, those things *of the type we are considering* which do not lie in S. As long as we know which universal set we are dealing with, no ambiguity arises.

Taking complements reverses inclusion relations between sets. If $S \subseteq T$ then $T' \subseteq S'$. This is because, if there are *more* things in T than in S, then there are *fewer* things not in T than there are not in S. It is obvious from a Venn diagram (Figure 32).

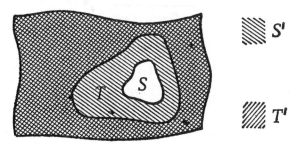

Figure 32

Taking complements of sets is closely related to negating statements. We can use set theory to solve certain kinds of logical problems. Consider this list of statements:

(i) Animals which cannot be seen at dusk are grey.

(ii) The neighbours do not like things that keep them awake.

(iii) Anything that sleeps heavily snores loudly.

(iv) The neighbours like animals which are visible at dusk.

(v) All elephants are heavy sleepers.

(vi) Anything that snores loudly wakes the neighbours.

We can turn these into set-theoretic statements, by putting

A = the set of things that wake the neighbours,

B = the set of things that sleep heavily,

C = the set of things that snore loudly,

D = the set of animals that are visible at dusk,

E = the set of elephants,

F = the set of things liked by the neighbours,

and G = the set of grey things.

Then statement (i) says that anything *not* in D is in G, that is,

$$\text{(i)} \qquad D' \subseteq G.$$

Similarly the other statements become

$$\text{(ii)} \qquad A \subseteq F'$$
$$\text{(iii)} \qquad B \subseteq C$$
$$\text{(iv)} \qquad D \subseteq F$$
$$\text{(v)} \qquad E \subseteq B$$
$$\text{(vi)} \qquad C \subseteq A.$$

Taking the complements of D and F, we can deduce from $D \subseteq F$ that

$$F' \subseteq D'.$$

Now we can string all the statements together:

$$E \subseteq B \subseteq C \subseteq A \subseteq F' \subseteq D' \subseteq G$$

and since a subset of a subset is a subset, these imply that

$$E \subseteq G,$$

i.e., *all elephants are grey.*

There is much more to the connection between set theory and logic. The idea was first thought of by George Boole (1815–64), and the consequent theory is known as *Boolean algebra*.[10]

Using complements we can explain a phenomenon which you may have noticed already. The various set-theoretic identities, or 'laws', seem to come in *pairs*. If I take a law involving the signs $\cup$ and $\cap$, and turn all the $\cup$s into $\cap$s and all the $\cap$s into $\cup$s, the result is another law. The laws mentioned in the section on unions and intersections were in fact written down in such pairs.

This is no accident. It is a consequence of two more identities, known as *De Morgan's Laws*: for any sets A and B, we have

$$(A \cup B)' = A' \cap B'$$
$$(A \cap B)' = A' \cup B'.$$

Even these come in pairs. Now anything not not in a set S is in S, and conversely; so $S'' = S$. So we can rewrite these as

$$A \cup B = (A' \cap B')'$$
$$A \cap B = (A' \cup B')'. \tag{†}$$

Take any set-theoretic law, such as

$$(A \cup B) \cap C = (A \cap C) \cup (B \cap C).$$

Change all the As, Bs, and Cs to their complements, obtaining

$$(A' \cup B') \cap C' = (A' \cap C') \cup (B' \cap C').$$

This is also a law, because the first equation is true for *any* sets A, B, and C. Now take complements of both sides:

$$((A' \cup B') \cap C')' = ((A' \cap C') \cup (B' \cap C'))'$$

and use De Morgan's laws, as rewritten in (†), to simplify. The left-hand side becomes

$$(A' \cup B')' \cup C$$

which on further application of (†) is

$$(A \cap B) \cup C.$$

(You must remember that $A'' = A$, $B'' = B$, $C'' = C$.) Similarly the right-hand side is

$$(A' \cap C')' \cap (B' \cap C')'$$

or

$$(A \cup C) \cap (B \cup C).$$

So we have shown that

$$(A \cap B) \cup C = (A \cup C) \cap (B \cup C),$$

which is the original law, with $\cup$s and $\cap$s interchanged. The same method works for any law which involves only unions and intersections.

In consequence, our work in proving theorems is halved: for every theorem we prove we get a second free of charge.

Geometry as Set Theory

Euclid makes an attempt to define certain basic geometrical objects, such as 'point' and 'line'. For instance, a point is supposed to be something that has position but no size. If you analyse the idea of 'position' it turns out to be as difficult to define as 'point'; and the two concepts chase each other round in circles.

Now any definition must start somewhere. The dictionary defines 'the' as '*the* definite article'.[11] If you didn't know what 'the' meant this would not be very helpful! Euclid tried to relate his idealized points and lines to objects in the physical world. Unfortunately nothing in the real world behaves exactly like his ideal objects. Even very small sub-atomic particles have some size. (Indeed, if quantum theory is right, the whole idea of size becomes hazy for very small distances: it is physically impossible to measure distances smaller than, say, a trillionth of a centimetre. To do so would require so much energy that you'd blow what you were measuring to bits.) One way round this is to take the basic ideas of point and line as *undefined* terms, and then state how you wish them to behave. This is the modern version of the axiomatic method, and I shall have more to say about it in Chapter 8.

The notion of a plane composed of individual points is an appealing one. Another way to make it logically sound is to define 'plane' and 'point' in terms of already known mathematical objects. We cannot define anything which *is* a plane in the physical sense, but we can define an object which behaves in the way that an idealized Euclidean plane should behave.

As we remarked in Chapter 2, the ideas of coordinate geometry allow us to label each point of the plane with a unique pair (x, y) of coordinates. We now have a mysterious object 'point' associated with a straightforward object 'pair of real numbers'. The straightforward object will do everything that we want the mysterious one to do. If we are unwilling to indulge in mysticism, we can *define* a point to be a pair (x, y) of real numbers. The plane is composed of all points; so we can *define* the plane to be the *set* of all pairs of real numbers.

What about lines? If you go back to coordinate geometry, you will find that a line consists of those points (x, y) that satisfy an equation of the form

$$ax + by = c$$

for fixed a, b, c, For example, $1 . x + (-1) . y = 0$ gives the diagonal line through the origin from bottom left to top right. We can define a line to be the set of all pairs (x, y) satisfying such an equation. In the same way, the equation determining a circle can be used to define a certain set of points, which corresponds to the geometrical idea of a circle.

A point lies on a line, geometrically, if it is a set-theoretic member of the line. So a point lies on two lines L and M if it is a member of L and a member of M, in other words, if it is a member of the intersection $L \cap M$. Set-theoretic intersection corresponds to geometrical intersection.

Proceeding in this way, using coordinate geometry as inspiration, you can set up the whole of Euclidean geometry as a part of set theory. From the way that you want geometry to behave, you can construct a purely mathematical theory. But now, instead of indulging in deep metaphysical arguments about the 'real' geometry, you can say: here is a mathematical theory. It deals with things which I call 'points' and 'lines'. I suspect that in the real world very small dots and very thin lines will behave in approximately the same way. And then people can go away and do experiments, to see if you are right. And even if it turns out that with very exact measurements you are wrong, you will still have a nice theory.

I now want to generalize the idea of pairs of numbers. It is important to notice that the pairs we used above are *ordered*, that is, the pair $(1, 3)$ is *not* the same as the pair $(3, 1)$. Draw them on a sheet of graph paper. (This is in contrast to the unordered pairs $\{1, 3\}$ and $\{3, 1\}$ provided by taking sets; these *are* equal, as we agreed earlier.)

Given any two sets A and B we can define[12] ordered pairs (a, b), where $a \in A$ and $b \in B$. 'Ordered' means that

$$(a, b) = (c, d)$$

if and only if $a = c$ *and* $b = d$. Then we can define the *Cartesian product*

$$A \times B$$

to be the set of all possible ordered pairs (a, b), where $a \in A$ and $b \in B$. (The name is in honour of Descartes, who invented coordinate geometry.)

Suppose that $A = \{\triangle, \square, \bigcirc\}$ and $B = \{£, \$\}$. We can picture $A \times B$ as shown in Figure 33.

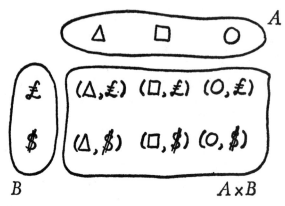

Figure 33

Notice that $A \times B$ is not the same set as $B \times A$. The latter contains, among others, the element ($£$, $\triangle$) which is not in $A \times B$.

If we recall the notation **R** for the set of real numbers, then the plane, as defined above, is the set **R** $\times$ **R**. It is customary to use the simpler notation $\mathbf{R}^2$. All of Euclidean geometry can be thought of as a study of the subsets of $\mathbf{R}^2$.

Chapter 5 What is a Function?

In elementary mathematics one comes across various objects designated by the term *function*: the logarithmic function, the trigonometric function, the exponential function. What these have in common is that for any number x the function takes a well-defined value, namely $\log(x)$, $\sin(x)$, $\cos(x)$, $\tan(x)$, e^x, and so on.

One also learns how to draw graphs representing functions, by

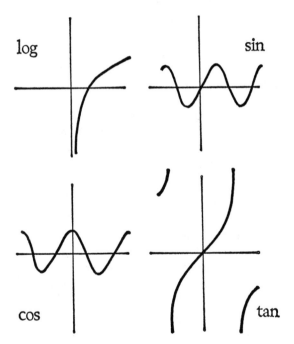

log   sin

cos   tan

Figure 34

plotting the value of the function at x against the value of x. Figure 34 illustrates this for four common functions.

In traditional terminology x is the *variable*; and the function assigns to each value of the variable x another value y. If the function is denoted by some symbol such as f, we write

$$y = f(x).$$

If f is the function 'log' then $y = \log(x)$; if f is the function 'sin' then $y = \sin(x)$.

Neither y nor x are functions. In fact it is very difficult to say exactly what they are. And $f(x)$ is not a function either, because it is the *value* of the function at x. It is f that is the function. The 'variables' x and y exist only to tell us what f does. The function 'square of' takes the value x^2 for any given value of x. This can be expressed briefly by the formula

$$y = x^2$$

but without being told in advance you would not be certain whether this formula was the definition of a function or an equation to be solved.

On Formulae

Most of the functions one meets in school mathematics can be defined by a formula: $y = x^2$, $y = \sqrt{x}$, $y = |x|$, or more complicated ones like

$$y = 7x^4 + \frac{\sin(x)}{1+x}.$$

This encourages a belief that mathematics *is* formulae; that the object of a mathematician's life is to produce more and more complicated formulae and to use these to perform more and more difficult calculations. This is not so. Worse, a *blind* manipulation of formulae, without understanding what you are doing, can lead you into making any number of silly errors. The example I want to give involves calculus, but it is not used in any essential way and anyone who doesn't know calculus should still be able to follow.

I have in the past set classes the task of differentiating the function

$$y = \log\left(\log\left(\sin\left(x\right)\right)\right).$$

If you follow blindly the standard rules of the calculus, you will obtain the answer

$$\frac{1}{\log (\sin (x))} \cdot \frac{1}{\sin (x)} \cdot \cos (x)$$

or

$$\frac{\cot (x)}{\log (\sin (x))}.$$

Most members of the class were happy with this. I then asked them to sketch the graph of $\log \big(\log (\sin (x))\big)$. This caused great consternation, because it revealed that the formula didn't make any sense. For any value of x, $\sin (x)$ is at most equal to 1, so $\log \big(\sin (x)\big) \leq 0$. Since logarithms of negative numbers cannot be defined, the value $\log \big(\log (\sin (x))\big)$ does not exist; the formula is a fraud.

On the other hand, the 'derivative' $\cot (x)/\log \big(\sin (x)\big)$ does make sense, for certain values of x; namely those where $\sin (x) > 0$.

Some people might enjoy living in a world where one can take a function which does nót exist, differentiate it, and end up with one that does exist. I am not one of them.

A given formula may not make sense for certain values of the variable x. Thus $1/x$ is not defined when $x = 0$, $\log (x)$ is not defined for $x \leq 0$, and $\tan (x)$ is not defined for x an odd multiple of $90°$. More complicated formulae can go wrong in more complicated ways; thus

$$\frac{\log (x^2 - 1)}{x^2 - 5x + 6}$$

is not defined if $-1 \leq x \leq 1$, or if $x = 2$, or if $x = 3$.

Again, there are many useful functions which are not easily definable using formulae. (A question which arises here is: which kinds of formulae? The function 'sine' is not definable by formulae unless you invent a new symbol 'sin'.) For many purposes in mathematics we need functions like the integer part $[x]$ defined by

$$[x] = \text{the largest integer} \leq x.$$

Or we might want a function such as that shown in Figure 35, defined by

$$f(x) = \begin{cases} (x+1)^2 & \text{if } x < -1, \\ 0 & \text{if } -1 \leq x \leq 1, \\ (x-1)^2 & \text{if } 1 < x. \end{cases}$$

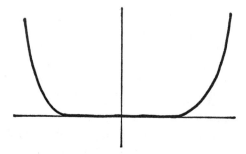

Figure 35

In the theory of Fourier analysis one encounters functions like the square wave (Figure 36).

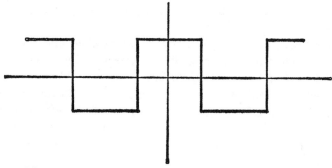

Figure 36

For years mathematicians debated whether or not this *was* a function. It didn't look like any of the familiar functions, and there seemed to be no formula for it. The problem got worse when Fourier showed that the infinite series

$$\sin (x) + \tfrac{1}{3} \sin (3x) + \tfrac{1}{5} \sin (5x) + \dots$$

could be summed to give a square-wave function. For now we have the nice, homely trigonometric functions giving birth to a weird creature with corners!

The ensuing squabble took a century or more to settle. In part this was because the problem 'Is this a function?' was jumbled up with others, like 'What is an infinite series?', but mostly it was because every mathematician had his own idea of what functions ought to look like, and couldn't agree with anybody else.

More General Functions

We have seen that a function f need not be defined for all values of the variable x. If $f(x)$ is given by a formula, that formula does not necessarily make sense for all x.

The values of x for which the function *is* defined form a subset of the set **R** of real numbers. This set is called the *domain* of f, and it tells to which values of x the function f can be applied.

From all the properties common to the examples of functions above, we single out one as being of overriding importance: *the value of* f(x) *is uniquely specified for every element* x *of the domain.*

As well as the domain, there is another set associated with a function, and known as its *range*. This consists of all the possible values that the function may take, when evaluated on elements of the domain. The range of the function 'sin' is the set of real numbers between -1 and 1. The range of the function 'square of' is the set of positive reals.

The range of even a simple function may be very complicated. The function f whose domain is the set of positive integers and which satisfies

$$f(x) = \sqrt{(x!)}$$

(positive square root) has as range the set of all square roots of factorials. It is hard to give a more helpful characterization than that!

For this reason we are less interested in the precise range of a given function. It is often more useful to have some description of where the values lie, of a simple nature. Any set T such that all values $f(x)$ lie in T will perform this role. Such a set T is called a *target* for f; and we say that f is a function *from* the domain D *into* the target T.

A function, then, consists of three things:

(1) a domain D,
(2) a target T,
(3) a rule which, for *every* $x \in D$, specifies a *unique* element $f(x)$ of T.

Item (3) is the heart of the matter.
It is important that $f(x)$ be *uniquely* defined, so that there is no

ambiguity attached to it. Taking square roots does not define a function unless we specify whether we want the positive or negative square root. It is also important that it be defined for *every x* in the domain, in order that knowledge of the domain may tell us when it is safe to use *f*. It is not particularly important to know the exact range of *f* – often this is very hard to work out, and we want to be able to use *f* without worrying about the problem – so we are free to choose *T* in as convenient a way as we like.

The only other term in (3) which requires explanation is the word 'rule'. For the moment I shall assume that we all know what a 'rule' is: it is a way of working out $f(x)$ given any particular x. But to this I should add that it is sufficient if $f(x)$ is *in principle* calculable from x. In *practice*, the computations may be too hard or take too long to be possible; they might depend upon solving some very difficult problem.

So far, the domain and target have always been sets of real numbers. But our conditions (1), (2), (3) make sense as long as *D* and *T* are *sets*. Furthermore, rules of the type envisaged in (3) crop up naturally in situations where *D* or *T* are not sets of real numbers. This remark is important for everything following, so I shall give several examples.

(i) Let *D* be the set of all circles, *T* the real numbers, and for any circle *x* define

$$f(x) = \text{the radius of } x.$$

(ii) Let *D* be the set of positive integers, *T* the set of all sets of prime numbers; for any $x \in D$ define

$$f(x) = \text{the set of prime factors of } x.$$

(iii) Let *D* be a subset of the plane, let *T* be the plane (thought of as the set $\mathbf{R}^2$), and for $x \in D$ let

$$f(x) = \text{the point 5 cm to the right of } x.$$

(iv) Let *D* be the set of all functions, *T* the set of all sets, and for any function *x* define

$$f(x) = \text{the domain of } x.$$

In each case the rule which defines $f(x)$ is unambiguous. Example (iii) is particularly interesting. In Chapter 1 we define a *transformation T* by

$$T(x, y) = (x+5, y).$$

This is the same rule that determines *f*; essentially there is no difference between *T* and *f*.

The modern concept of 'function' is tailored to suit all of these examples. From now on a *function* will be anything that satisfies conditions (1), (2), (3), where *D* and *T* may be quite general *sets*. Our earlier functions are then a special kind, namely those whose domain and target lie inside the set of real numbers.

The function *f* with $f(x, y) = (x+5, y)$ is an example of what, in calculus, is known as a *function of two variables*. So functions of two variables also come under our general heading; for then the domain will be a set of *pairs* (x, y) of real numbers, or a subset of $\mathbf{R}^2$.

The function concept is a strong contender for the most important in contemporary mathematics, because it has such a wide range of applications. The idea of a function will turn up again and again as we proceed, in many different guises. For this reason, it is worth developing a few general notions about them.

Properties of Functions

If the domain and target are not subsets of **R**, it is not possible to draw a graph of the function. Indeed the graphical representation is not a very helpful picture for our generalized function concept. A better way to think of functions is shown in Figure 37.

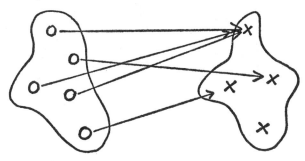

Figure 37

The arrows in this picture represent the 'rule' which tells us what $f(x)$ is.

The standard notation expressing the fact that *f* is a function with domain *D* and target *T* is

$$f: D \rightarrow T,$$

which uses the arrow in the same way.

In Figure 37 there is an element of *T* to which no arrow points. The range of *f* is not the whole of *T*. If the range of *f* is the whole of *T*, then *f* is said to be a function *onto*[1] *T*. Another common word for such a function is *surjection* (from the Latin: *f* throws *D* on to *T*). To put it in pictures, *f* is a surjection provided every element of *T* can be reached by going along some arrow, as in Figure 38.

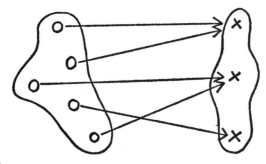

Figure 38

It does not matter if, as is the case in Figure 38, certain elements of *T* lie at the end of more than one arrow. If every element of *T* lies on at most one arrow (perhaps on none) then *f* is an *injection*. Injections need not be surjective (Figure 39).

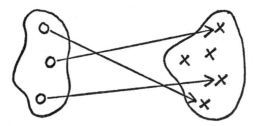

Figure 39

If we have a function *f*: *D* → *T* which is both an injection and a surjection, then the arrows pair off elements of *D* and *T*: an element of D at the tail end, an element of T at the head. No two elements of D are paired with the same element of T, because f is an injection. No two elements of T are paired with the same element of D, because of the uniqueness clause in part (iii) of the definition of function. Every element of D occurs, because D is the domain; every element of T occurs because f is a surjection. Although it is not apparent from the notation, the situation is perfectly symmetrical; and if we turn all the arrows round we define another function

$$g : T \to D$$

in the opposite direction. And g is also both an injection and a surjection. (See Figure 40.)

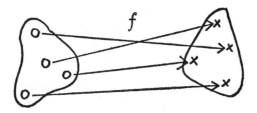

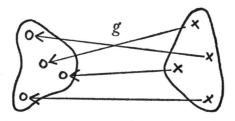

Figure 40

Functions which can be turned round in this way will play a prominent part in some of our later investigations. They are known as *bijections*, or as *one-to-one correspondences*.

There is nothing to stop you turning round the arrows if f is not a bijection. But you won't get a function. If f is not injective then, on turning round, some element of T will be at the tail end of two different arrows, and the reversed 'function' is not uniquely defined. If *f* is not surjective, there will be elements of *T* for which the reversed function is not defined at all.

In Chapter 1 we combined transformations *F* and *G* to give a new transformation *FG*, corresponding to the idea 'do *G* and then do *F*'. But transformations are a kind of function. Can we combine functions in the same way?

Suppose we take two functions *f* and *g*, and try to define a function *fg*. As for transformation, we want

$$fg(x) = f(g(x))$$

for the relevant *x*.

For this formula to make sense, several conditions must hold. We can't work out $f(g(x))$ unless $g(x)$ is defined, so

(i) *x* must lie in the domain of *g*.

Then, to work out $f(g(x))$, we need to know that

(ii) $g(x)$ lies in the domain of *f*.

Suppose then that $f: A \rightarrow B$ and $g: C \rightarrow D$. The most we can hope for is that *fg* has domain *C*, because of (i). In order to define it on all of *C*, (ii) must hold for all $x \in C$. In other words, the range of *g* must lie inside the domain *A* of *f*. If this condition holds then *fg*, defined by the formula, will be a function from *C* into *B*. This is illustrated in Figure 41.

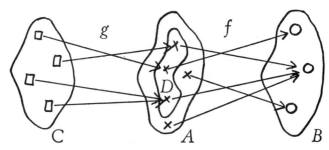

Figure 41

The function *fg* corresponds to the idea 'do *g*, then do *f*'. If we have three functions *f*, *g*, and *h*, and if their ranges and domains fit together properly, we could do all three in succession:

first *h*, then *g*, then *f*. There are two ways of accomplishing this by combining functions in pairs. Either do *h* and then do *fg*, or do *gh* and then do *f*. These correspond to the two expressions

$$(fg)h \qquad f(gh).$$

Fortunately these are always equal. It makes no difference in the end if we do '*h* and then *g*-and-then-*f*' or if we do '*h*-and-then-*g* and then *f*'. In pictures, we have Figure 42.

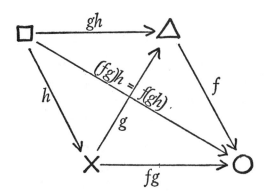

Figure 42

Or instead we could calculate:

$$(fg)h(x) = (fg)(h(x)) = f(g(h(x)))$$
$$f(gh)(x) = f(gh(x)) \quad = f(g(h(x))).$$

Either way, we can see that always

$$(fg)h = f(gh).$$

We say that our method of combining functions satisfies the *associative law*.

I said above the we could combine *f*, *g*, and *h*, provided their ranges and domains fit together properly. It is easy to see what this means: the range of *h* must be a subset of the domain of *g*, and the range of *g* must be a subset of the domain of *f*. Let's go back to the fraudulent formula

$$\log\left(\log(\sin(x))\right).$$

This is obtained from a combination of the functions sin, log, and log again. If we took *h* = sin and *f* = *g* = log above we would have $\log\left(\log(\sin(x))\right) = fgh(x)$.

The domain of sin is the whole of **R**; its range is the set of reals between −1 and 1. The domain of log is the set of positive reals; its range is the whole of **R**. The conditions which allow us to combine the functions are violated in several places: the range of log is not contained in the domain of sin, and the range of log is not contained in the domain of log. It is no wonder that the formula did not make sense!

Finally, let us return to the idea of reversing arrows, and make it mathematically respectable.

On any set D there is a function called the *identity* function, denoted by 1_D. It has domain D, range D, and is defined by

$$1_D(x) = x$$

for all $x \in D$. The effect of the function is to leave things unchanged. Not a very useful function? Assuredly, it is not a very *difficult* function. But it will be useful precisely when we wish to express the fact that a certain combination of functions leaves things unchanged.

We had, earlier, a bijection $f : D \to T$. Turning the arrows round gave another function $g : T \to D$. To express this symbolically, we note that everything is left unchanged by doing first g and then f: it travels up and down the same arrow. So

$$fg = 1_T.$$

Similarly   (†)

$$gf = 1_D.$$

The equations (†) are a symbolic expression of the fact that f and g can be obtained from each other by reversing arrows. We can say that f is the *inverse function* to g (and that g is inverse to f). The word 'the' is permissible here because it is easy to prove that inverses are unique: there is only one way to turn arrows round.

Summary

This chapter has been a little technical. The important points to remember for the rest of the book are:

> A function is defined on some set.
> It takes values in some set.
> It is defined if we know a rule which tells us how to find its

value for a given element, and if this value is uniquely specified.

Bijections, or one-to-one correspondences, are exactly those functions which have inverse functions.

I have not expanded here on the word 'rule': I have relegated to the notes[2] a more abstruse definition of function in terms of set theory, because it is of purely technical interest.

Chapter 6   The Beginnings of Abstract Algebra

Early on in algebra you are asked to simplify algebraic expressions, such as $2x+(y-x)$. Later on you get so used to doing this that you just look at the expression, and the answer $x+y$ presents itself without further thought.

Familiarity breeds contempt. You forget just how many painful stages you had to go through, how many different ideas you had to grasp, before this facility was acquired. If you try to write out in detail the steps required to perform the simplification, you will find that there are quite a lot of them. The way I simplified it – 'analysed out' – was this:

$$2x+(y-x) = 2x+(y+(-x)) \qquad (1)$$
$$= 2x+((-x)+y) \qquad (2)$$
$$= (2x+(-x))+y \qquad (3)$$
$$= (2x+(-1)x)+y \qquad (4)$$
$$= (2+(-1))x+y \qquad (5)$$
$$= 1.x+y \qquad (6)$$
$$= x+y. \qquad (7)$$

Steps (1) and (4) are minor, amounting to the use of definitions of $-x$ or of $y-x$, and step (6) is just arithmetic. But each ot the remaining steps uses the truth of certain general arithmetical laws – perhaps it would be better to say algebraic laws. At step (2) I assume that $a+b = b+a$. At (3) I use the law $a+(b+c) = (a+b)+c$. Step (5) uses the law $ax+bx = (a+b)x$, and step (6) the law $1.x = x$.

For the moment setting aside the laws relating to division, we can list the more important ones:

    (1) The associative law of addition:
$$(a+b)+c = a+(b+c).$$
    (2) The commutative law of addition:
$$(a+b) = (b+a)$$
    (3) The existence of zero:

There is a number 0 such that $a+0 = a = 0+a$ for any number a.

(4) The existence of additive inverses:

For any number a there is a number $-a$ such that $a+(-a) = 0 = (-a)+a$.

(5) The associative law of multiplication:
$$(ab)c = a(bc).$$

(6) The commutative law of multiplication:
$$ab = ba.$$

(7) The existence of unity:

There is a number 1 such that for any number a, $1a = a1 = a$.

(8) The distributive laws:
$$a(b+c) = ab+ac$$
$$(a+b)c = ac+bc.$$

Although there are a lot of laws, this does not make algebra complicated. In fact, the more laws there are, the easier things become, because we have more ways to simplify expressions.

Even some of our algebraic *notation* relies on the truth of some of the laws. The only reason we can write
$$a+b+c$$
unambiguously is because the associative law holds.

Most of elementary algebra consists of using these laws to prove formulae (although it is not always presented in such a light). The formula
$$(x+y)^2 = x^2+2xy+y^2$$
can be derived as follows. First note that for any number a we *define* a^2 to be $a.a$, and $2a$ to be $a+a$. Second, note that $a+b+c$ is a shortened form of $(a+b)+c$. Now proceed:

$$
\begin{aligned}
(x+y)^2 &= (x+y)(x+y) && \text{(notation)} \\
&= \big(x(x+y)\big)+\big(y(x+y)\big) && \text{(law 8)} \\
&= (xx+xy)+(yx+yy) && \text{(law 8)} \\
&= (x^2+xy)+(yx+y^2) && \text{(notation)} \\
&= (x^2+xy)+(xy+y^2) && \text{(law 6)} \\
&= \big((x^2+xy)+xy\big)+y^2 && \text{(law 1)} \\
&= \big(x^2+(xy+xy)\big)+y^2 && \text{(law 1)} \\
&= (x^2+2xy)+y^2 && \text{(notation)} \\
&= x^2+2xy+y^2 && \text{(notation)}.
\end{aligned}
$$

With a little more work, you could prove that the usual expansions of $(x+y)^3$, $(x+y)^4$, hold good; or even give a proof of the binomial theorem (for integer powers); all using only laws (1)–(8).

Rings and Fields

The ordinary systems of numbers ($\mathbf{Z}$, $\mathbf{Q}$, and $\mathbf{R}$) are not the only ones in which the laws (1)–(8) hold. They also hold (though I won't give a proof yet) for the integers modulo 6, for example. To take a few instances:

$$(1+4)+3 = 5+3 = 2 = 1+7 = 1+(4+3)$$
$$2.5 = 4 = 5.2$$
$$1.4 = 4 = 4.1$$
$$3(2+5) = 3.1 = 3 = 0+3 = (3.2)+(3.5).$$

In consequence, the formula for $(x+y)^2$ will also hold for the integers modulo 6, because only laws (1)–(8) were used to derive it.

There is nothing special about 6 here. The integers modulo 2, 3, 4, 5, 6, 7, ..., in fact modulo n for any n, also satisfy (1)–(8). The formula for $(x+y)^2$ holds for these systems too, and has the same proof.

Mathematicians are lazy creatures at heart. The labour of writing out a proof of the formula for each system of integers mod 2, mod 3, mod 4, mod 5, ..., seems too great to justify the results, particularly when the proof is the same every time. Why not agree that the proof will work for any system which satisfies (1)–(8)? To make matters clearer, why not give a name to such systems, so that we can avoid listing all eight laws every time?

The name currently in vogue is: *commutative ring with unity*. This is a trifle cumbersome. A *ring* is any set S having two operations $+$ and $.$ defined on it, such that if s and t lie in S then $s+t$ and $s.t$ also lie in S, and such that laws (1)–(5) and (8) hold. (Here st is to be interpreted as $s.t$.) The ring is *commutative* if (6) also holds. It has a *unity* if (7) holds. The shortest name 'ring' is used for the object most frequently encountered. But in this book we won't encounter any non-commutative rings, because our source of examples is too limited.

The use of symbols $x+y$ and xy for 'addition' and 'multiplication' in the ring is just a convention – though a useful one! If instead we had used $\square$ and $\circ$, we would want the corresponding laws to hold: law (8), in this form, would read

$$a \circ (b \square c) = (a \circ b) \square (a \circ c)$$
$$(a \square b) \circ c = (a \circ c) \square (b \circ c).$$

The underlying set S of a ring need not be a set of numbers. Even for the integers mod 7, where S is the set $\{1, 2, 3, 4, 5, 6, 0\}$, the elements are not really numbers; in fact it doesn't matter what they are as long as we use the tables on pages 31 and 32 to define addition and multiplication. Again, we might take *any* set T, and put

$$S = \{\text{all subsets of } T\}.$$

We define, for a and $b \in S$.

$$a+b = (a \cup b)-(a \cap b)$$
$$ab = a \cap b.$$

(See Figure 43.)

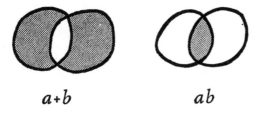

$$a+b \qquad\qquad ab$$

Figure 43

It is now a long but elementary exercise in set theory to show that laws (1)–(8) hold for these operations on S. The empty set $\emptyset$ plays the part of 0 in law (3), while T does for 1 in law (7). The two sides of (1) are both represented by Figure 44.

What will x^2 be in this ring? Recall that $x^2 = xx$. The elements of xx are those which lie in both x and in x: in other words, those in x. So $xx = x$. The ring has the curious property that $x^2 = x$ for *every* element! If we took T to be a set with n elements, then S has 2^n elements; and we have a ring in which the quadratic equation

$$x^2-x = 0$$

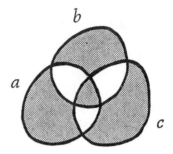

Figure 44

has 2^n solutions. If T is an infinite set, then the equation has infinitely many solutions!

We said that in *any* ring it is always true that $(x+y)^2 = x^2+2xy+y^2$. If every element satisfies $x^2 = x$ this reduces to

$$x+y = x+2xy+y$$

from which – using, incidentally, laws (4), (1), and (3) – it follows that

$$2xy = 0$$

for any x, y. Actually, more is true. For any element x of S we have

$$2x = x+x = (x \cup x)-(x \cap x) = x-x = \emptyset = 0.$$

So certainly $2xy = 0$. Although the ring has several very peculiar properties, the formula for $(x+y)^2$ does not lead to any inconsistencies.

The branch of mathematics known as ring theory consists of deductions from laws (1)–(5) and (8): those theorems which are true of all rings. If in the course of his work a mathematician chances upon a system satisfying all these laws he says 'Aha! a ring!' and knows that from this observation a certain body of standard properties will follow. (Seldom does this solve all his problems.)

If we bring in division, two more laws become important.

(9) The existence of multiplicative inverses:
   If $a \neq 0$ there exists an element a^{-1} such that $aa^{-1} = 1 = a^{-1}a$.
(10) $0 \neq 1$. (This is just to exclude certain trivial systems.)

A set S having operations of addition and multiplication which satisfy laws (1)–(10) is called a *field*. The results of Chapter 3 about reciprocals show that the integers modulo n form a field if and only if n is prime. There are therefore plenty of rings which are not fields; for example the integers mod n when n is *not* prime.

Historically the notions 'ring' and 'field' arose from the study of algebraic numbers: numbers satisfying a polynomial equation, such as $x^2 - 2 = 0$, or $17x^{23} - 5x^5 + 439 = 0$. The first is satisfied by $\pm\sqrt{2}$ (I have no idea about the second!) and at a certain stage in the theory it is helpful to look at all the numbers $a + b\sqrt{2}$ for integers a and b. Because

$$(a+b\sqrt{2})(c+d\sqrt{2}) = (ac+2bd)+(ad+bc)\sqrt{2}$$

it turns out that these form a ring.

If instead we allow a and b to be rational then we can find inverses:

$$(a+b\sqrt{2})^{-1} = \left(\frac{a}{a^2-2b^2}\right) + \left(\frac{-b}{a^2-2b^2}\right)\sqrt{2}.$$

So now we have a field. Deep properties of algebraic numbers have been found by using ring theory and field theory. In particular this applies to the modern treatment of the insolubility by radicals of the general equation of degree 5.[1]

Application to Geometrical Constructions

A proper study of the quintic equation would take us too far afield, but we can illustrate some of the ideas involved on a problem which uses less machinery.

A famous problem of Greek geometry (often cast in the form of a legend involving an oracle at Delos[2]) asks for a geometrical construction of a line of length $\sqrt[3]{2}$, given a line of length 1. The construction is to be performed under the Platonic restriction to ruler and compasses alone. The Greeks could not find a solution (although they *did* find one using conic sections).

We will show that no such construction exists.

Given lines of lengths r and s it is possible to construct lines of lengths $r+s$, $r-s$, rs, and r/s by the methods shown in Figure 45. (We assume a line of length 1 is given, to fix the 'scale'.)

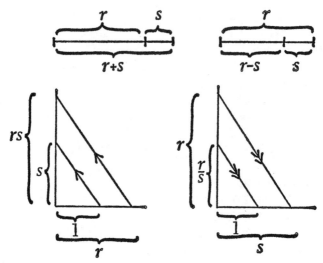

Figure 45

The set **K** of constructible lengths is a subset of the set **R** of real numbers. We have just seen that in **K** we can add, subtract, multiply, and divide; from which it is easy to show that **K** is a *field*. We can say that **K** is a *subfield* of **R**.

Another thing we can do inside **K** is take square roots of positive lengths. This follows from Figure 46.

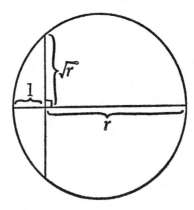

Figure 46

If we start with a given length 1 then using the constructions of Figure 45 we can construct lengths 2, 3, 4, ... $\frac{1}{2}$, $\frac{1}{3}$, $\frac{2}{3}$, ..., and in general all *rational* lengths. Then, for any rational number r, we can construct $\sqrt{r}$, as in Figure 46. Then we can get all the lengths of the form

$$p + q\sqrt{r}$$

for rational p and q. The set of all such numbers forms a field, call it F_1, because we can define inverses by

$$(p+q\sqrt{r})^{-1} = \left(\frac{p}{p^2-rq^2}\right) + \left(\frac{-q}{p^2-rq^2}\right)\sqrt{r},$$

a formula generalizing the one given for $p + q\sqrt{2}$.

Now we can start again. Pick an element $s \in F_1$, construct $\sqrt{s}$; then construct all numbers $p + q\sqrt{s}$ for p, q belonging to F_1. This gives a larger field F_2. Repeating the process on F_2 we get a field F_3. In general we get an increasing series of fields:

$$\mathbf{Q} \subseteq F_1 \subseteq F_2 \subseteq F_3 \subseteq \ldots \subseteq F_k \subseteq F_{k+1} \subseteq \ldots$$

and we can construct any length in any F_i.

Are there any other constructible lengths? Of course, we could pick different elements $r, s, \ldots$ when taking square roots. But this still leads to a similar series of fields. Are there any not obtainable by our procedure?

Any geometrical construction can be broken down into a sequence of steps of three kinds:

(i) finding the point of intersection of two lines, whose end points are already constructed;

(ii) finding the point(s) of intersection of a line and a circle, where the ends of the line, the centre of the circle, and a line of length equal to the radius of the circle, are already constructed;

(iii) finding the point(s) of intersection of two circles, whose centres are already constructed, and whose radii are equal to lengths already constructed.

Analysing these by coordinate geometry, one finds that step (i) only brings in lengths which can be obtained from those already found by addition, subtraction, multiplication, and division. Steps (ii) and (iii) bring in square roots of known lengths as well, but nothing more. So *every* constructible length lies in one of our fields F_i, for suitable choices of $r, s, \ldots$ when we take square roots.

Now we come to the problem of constructing $\sqrt[3]{2}$. If this is possible, then $\sqrt[3]{2}$ must lie in one of the F_i. Now it does not *look* as if $\sqrt[3]{2}$ can be expressed using only square roots: but perhaps appearances are deceptive. Are you sure that there may not be some complicated expression, perhaps

$$3 + \tfrac{2}{7}\sqrt{(5 + 6\sqrt{7})} - \sqrt{13}$$

which is equal to $\sqrt[3]{2}$?

This certainly seems unlikely, because of the *cube* root. Cube roots and square roots are quite different beasts. This difference must be exploited.

First we prove that $\sqrt[3]{2}$ is not a rational number. The proof is a variation on a standard proof that $\sqrt{2}$ is not a rational number.

Suppose, on the contrary, that $\sqrt[3]{2}$ is rational, so that there are integers c, d with

$$\sqrt[3]{2} = c/d.$$

It is possible that c and d have some common factors: if so we cancel these out. Either way we can find integers e, f *without common factors* such that

$$\sqrt[3]{2} = e/f.$$

Cubing and multiplying by f^3 gives us

$$2f^3 = e^3.$$

Therefore e^3 is *even*. The cube of an odd number is odd, so e cannot be odd: therefore e *is even*. So there is an integer g such that

$$e = 2g.$$

Then

$$2f^3 = e^3 = (2g)^3 = 8g^3$$

so that

$$f^3 = 4g^3.$$

Therefore f^3 is even, and by a similar argument f *is even*. So there is an integer h such that

$$f = 2h.$$

But now we see that e ($=2g$) and f ($=2h$) have the common factor 2. But e and f *do not have* common factors!

If e and f exist, they have self-contradictory properties. Therefore they do not exist. Therefore c and d don't exist: and it follows that $\sqrt[3]{2}$ is not a rational number.

Assume, for argument's sake, that $\sqrt[3]{2}$ *can* be constructed. We have just seen that it isn't rational. So it must lie in some field F_k, with corresponding choice of elements r, s, ... There is no harm in taking this k to be as *small* as possible.

Let us write $x = \sqrt[3]{2}$. Since $x \in F_k$ it follows that
$$x = p + q\sqrt{t} \tag{†}$$
where p, q, and t lie in F_{k-1}, but $\sqrt{t}$ does not. (If $\sqrt{t}$ did then $F_k = F_{k-1}$, and so $x \in F_{k-1}$, contradicting our choice of the smallest k.) Now x satisfies
$$x^3 - 2 = 0.$$
Substituting from (†) we find that
$$a + b\sqrt{t} = 0$$
where
$$a = p^3 + 3pq^2t - 2 \qquad b = 3p^2q + q^3t.$$
This means that a and b must both be zero. Because if $b \neq 0$ we would have
$$\sqrt{t} = -a/b$$
which lies in F_{k-1}. But we have just said that $\sqrt{t}$ does *not* lie in F_{k-1}. So $b = 0$. Consequently $a = 0$ as well.

Now consider the number
$$y = p - q\sqrt{t}.$$
You will find that
$$y^3 - 2 = a - b\sqrt{t}$$
with a and b as above. But these are both 0, so that
$$y^3 - 2 = 0 \text{ also.}$$
This means that y is another cube root of 2. Now x and y are both real numbers; and there is only *one* real cube root of 2. The only possibility left is that $x = y$; and even this is not much of a possibility, because then
$$p + q\sqrt{t} = p - q\sqrt{t}$$
so that $q = 0$. From (†) we then have $x = p$, but p lies in F_{k-1}. So x lies in F_{k-1}. But again this contradicts our explicit choice of the *smallest* possible k.

With meticulous logic we have crashed headlong into a contradiction. The only dubious item is our assumption that $x = \sqrt[3]{2}$ *can* be constructed, so this must be the source of our troubles: we can avoid a contradiction only if $\sqrt[3]{2}$ *cannot* be constructed. This must therefore be the case.

Other problems about geometrical constructions may be solved in similar ways. Trisecting the angle 60° boils down to constructing x such that

$$x^3 - 3x = 1$$

and a similar argument proves this impossible. 'Squaring the circle' implies constructing $x = \pi$. Because the numbers in F_i are obtained by taking only square roots, it can be seen that they must satisfy some polynomial equation

$$a_n x^n + a_{n-1} x^{n-1} + \ldots + a_1 x + a_0 = 0$$

whose coefficients a_i are all rational. A famous and difficult theorem of Lindemann[3] asserts that π satisfies no such equation. So the circle cannot be squared.

In this connection the problem of constructing regular polygons should be mentioned. The construction of the n-sided regular polygon is intimately bound up with the polynomial equation

$$x^{n-1} + x^{n-2} + \ldots + x + 1 = 0.$$

A deep analysis of this shows that the regular n-gon can be constructed if and only if

$$n = 2^a . p_1 \ldots p_b$$

where the ps are *distinct odd* primes of the form

$$2^{2^c} + 1.$$

The only known primes of this form are for $c = 0, 1, 2, 3, 4$, when we get 3, 5, 17, 257, 65 537. Regular polygons with these numbers of sides may be constructed: this is very surprising in the cases 17, 257, or 65 537!

All these remarks refer to theoretically *exact* constructions. For practical purposes all one ever needs is sufficiently good *approximate* constructions, which exist for all reasonably small n; that is, small enough to be able to see the sides of the resulting polygon!

Congruences Again

I promised to prove that arithmetic mod n defines a ring.

Before I can even start, there is a snag: the laws of a ring demand equalities, and all I can as yet offer is congruences. Once I have got over this, there remains the problem of proving that the laws are true.

Let us, as usual, take the special case of arithmetic modulo 7. In Chapter 3 we found that the set $\mathbf{Z}$ of integers split up into seven subsets, corresponding to the seven days of the week. Wednesday, for example, corresponded to the set of numbers $\{\ldots -11, -4, 3, 10, 17\ldots\}$, where the numbers occurring are those of the form $7n+3$.

These are precisely the numbers *congruent* to 3 (mod 7). If we let $[x]$ denote the *set* of numbers congruent to x, then the sets of numbers corresponding to the days of the week are [0], [1], [2], [3], [4], [5], [6]. The sets $[x]$ are known as *congruence classes*. Furthermore, we have [7] = [0], [8] = [1], [9] = [2], etc., because a number is congruent to 7 if and only if it is congruent to 0.

In Chapter 3 we had 'equations' like $7 = 0, 8 = 1, 9 = 2$; but we did note that ' $=$ ' wasn't *really* equality. But for the equation [7] = [0] it *is* equality; the sets [7] and [0] are identical. This suggests that by putting square brackets everywhere we can recover equalities from congruences.

To do this will involve defining sums and products of congruence classes. This may seem daring. On the other hand, earlier in this chapter we defined them for the subsets of a fixed set T, so the idea of adding or multiplying sets is not entirely unheralded.

We can use the addition and multiplication tables mod 7 to define them: just put square brackets round everything. Then we will have

$$[4]+[5] = [2]$$
$$[3]+[1] = [4]$$
$$[5]\times[2] = [3]$$

and so on.

This is a step in the right direction, but it conceals an important simplification. Notice that [2] is exactly the same set as [9]. So [4]+[5] = [9]. And [3] = [10], so that $[5]\times[2] = [10]$. And we get back to ordinary arithmetic. In general, we have

$$[a]+[b] = [a+b]$$
$$[a]\times[b] = [ab]. \tag{$\ddagger$}$$

You should verify that these give the same tables or addition and multiplication as we had before. If you do, you will see that even the calculations involved in the two cases are basically the same: one is with brackets, the other without.

Have we come all this way just to end up with ordinary arithmetic again? Fortunately not. The laws of addition and multiplication are the same as for ordinary arithmetic, but we have some extra properties, such as $[7] = [0]$. We have ordinary arithmetic, *plus* the option of discarding multiples of 7. And this is exactly what arithmetic mod 7 ought to look like.

Since we now know that the laws of arithmetic mod 7 are basically the same as those of ordinary arithmetic, it is not surprising that we get a ring. To prove law (8), we proceed this way:

$$[a].([b]+[c]) = [a].([b+c]) \qquad \text{(definition of +)}$$
$$= [a(b+c)] \qquad \text{(definition of .)}$$
$$= [ab+ac] \qquad \text{(law 8 for the}$$
$$\text{ordinary integers)}$$
$$= [ab] + [ac] \qquad \text{(definition of +)}$$
$$= ([a].[b])+([a].[c]) \qquad \text{(definition of .)}$$

and all the other laws are equally easy. Everything is referred back to the ordinary integers.

The same ideas work modulo n for any n. First define congruences classes $[x]$, then define addition and multiplication using ($\ddagger$). Then prove laws (1)–(8).

It must be mentioned here that ($\ddagger$) is more subtle than it may appear. It tells us that $[1]+[3] = [4]$. *But it also tells us that* $[8]+[10] = [18]$. And since $[1] = [8]$ and $[3] = [10]$, it comes dangerously close to telling us two different things for the sum. But – and here one should breathe a sigh of relief – $[4] = [18]$, so it gives just one answer after all.

We are not always so lucky. If we split $\mathbf{Z}$ into two subsets P and Q:

$$P = \{\text{integers} \leq 0\}$$
$$Q = \{\text{integers} > 0\}$$

and let $[x]$ be whichever of P and Q the integer x belongs to – which is analogous to putting $[x]$ for whichever of the congruence classes x belongs to – we run into trouble. Using the formulae ($\ddagger$) to define $P+Q$ we get:

$$P+Q = [-5]+[1] = [-5+1] = [-4] = P$$
$$P+Q = [-3]+[6] = [-3+6] = [3] = Q.$$

This makes (‡) a pretty useless definition as far as P and Q are concerned! In fact, it isn't a definition at all.

However, for congruence classes (‡) is unambiguous, as can be proved without much effort. If $[a] = [a']$ and $[b] = [b']$, then $a - a' = jn$ and $b - b' = kn$, where j and k are integers and n is the modulus. So $(a+b)-(a'+b') = (j+k)n$, whence $[a+b] = [a'+b']$ and everything is all right. The operations $+$ and $.$ are *well-defined* by (‡) for congruence classes.

An Approach to Complex Numbers

The complex numbers[4] arise if we wish to solve the equation $x^2 + 1 = 0$. We introduce a new number i, defined so that $i^2 = -1$. In order to be able to add and multiply we must have numbers of the form $a + bi$ for real a and b. Finally we observe that if the laws of arithmetic are assumed to hold, nothing seems to go wrong. And as a bonus, we can divide as well.

This is all very fine, but it doesn't explain very much. It doesn't even *prove* that the laws *do* hold. And the number i can seem mysterious: which is why the real numbers are called 'real' and the imaginary numbers 'imaginary'. This is a great pity – not so much because it is a slight on the imaginaries, but because it lends the real numbers an air of respectability that they by no means deserve!

There is a way of introducing complex numbers which places them on a par with integers mod n. In the integers mod 7 we want the equation $7 = 0$ to hold, so we take congruences mod 7. In the complex numbers, we want $x^2 + 1 = 0$ to hold, so we take congruences mod $x^2 + 1$. At least, that is the general idea. First we must find somewhere to take them.

The somewhere must contain x. We want the reals to come into the final result, so we must put them in to start with. We want to do arithmetic, so we want things like $x + x$, xxx, $xxxx + 7x - 3$, and the like. These look like – indeed are – polynomials in x, with real coefficients. We already know how to add, subtract, or multiply polynomials; and we know that the laws of arithmetic hold. At least, we have always assumed that they hold, which is not quite the same thing. But we can prove that they do if we want. This means that the polynomials form a ring. We denote this ring by $R[x]$, where the R says the coefficients are real numbers, the x tells us what the variable is, and the square brackets have nothing whatever to do with congruences.

In the ring $R[x]$ we can take congruences to the modulus x^2+1. We say that two polynomials are congruent if their difference is divisible by x^2+1. Every polynomial is congruent to its remainder on division by x^2+1. For instance,
$$x^3+x^2-2x+3 = (x^2+1)(x+1)+(-3x+2)$$
so that
$$x^3+x^2-2x+3 \equiv -3x+2 \qquad (\mathrm{mod}\ x^2+1).$$
Indeed, every polynomial is congruent to a unique polynomial of the form $ax+b$, where a and b are real. We can eliminate all the terms of higher degree by subtracting multiples of x^2+1.

The constant polynomials look much like the real numbers, even taking congruences mod x^2+1. The polynomial x satisfies
$$x^2 \equiv -1 \qquad (\mathrm{mod}\ x^2+1)$$
and so behaves like the imaginary number i. The polynomials $ax+b$ then behave in the way we want our complex numbers $ai+b$ to behave.

In addition, we can prove that laws (1)–(8) hold for congruence classes (mod x^2+1) in exactly the way we proved them mod n; except that this time they all refer back to the polynomial ring $R[x]$.

Finally, we remark that
$$(ax+b)(-ax+b) \equiv -a^2x^2+b^2$$
$$\equiv a^2+b^2$$
so that we can find an inverse
$$\left(\frac{-a}{a^2+b^2}\right)x+\left(\frac{b}{a^2+b^2}\right)$$
for $ax+b$ whenever $ax+b \neq 0$. We have a field.

This is a bonus, because we started with $R[x]$, which isn't a field. But the same thing happened modulo n. We started with Z, which isn't a field. When n was a prime number, we discovered that the integers mod n formed a field. And much the same is happening here: the polynomial x^2+1 is 'prime' in the polynomial ring; it cannot be factorized.

We could carry on from here and develop all of the standard properties of complex numbers.

I wouldn't recommend this as the *best* way of introducing complex numbers to a class. But *if* they have a thorough grounding in arithmetic mod n, and *if* they have done some work on complex numbers, it is an illuminating parallel. We can even say what a complex number is: a congruence class of polynomials to the modulus $x^2 + 1$. A trifle offbeat, but certainly not mysterious.

In a Lighter Vein

The theory of rings and fields can be useful in circumstances far removed from abstract algebra.

The game of *solitaire* as you probably know, is played on a board with holes in the pattern

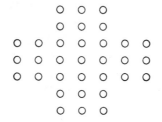

At the start, each hole except the centre contains a peg. The player may jump any peg over another horizontally or vertically adjacent peg into an empty hole, removing the peg jumped over. Diagonal moves are not allowed. The aim is to remove all pegs but one. Usually it is required that the last peg should end up in the centre. Anyone playing the game will observe, if he plays it enough, that although the final peg need not end up in the centre, it does not seem possible to make it finish anywhere. There is a limited number of final positions.

We shall ask: in what positions is it possible for the last peg to be? And we shall answer it by a method due to de Bruijn,[5] using a certain field with 4 elements. The elements will be 0, 1, p, and q: and addition and multiplication are defined by the tables

+	0	1	p	q
0	0	1	p	q
1	1	0	q	p
p	p	q	0	1
q	q	p	1	0

×	0	1	p	q
0	0	0	0	0
1	0	1	p	q
p	0	p	q	1
q	0	q	1	p

We shall not verify here that these tables do define a field, but it's true. Try a few calculations if you aren't convinced.

We observe that the equation

$$p^2+p+1 = 0 \qquad (\S)$$

holds. (This is crucial, and it is why we use the field chosen.) For

$$p^2+p+1 = q+p+1$$
$$= 1+1$$
$$= 0.$$

We assign integer coordinates to the holes on the board, like this:

$$
\begin{array}{ccccccc}
& & (-1,3) & (0,3) & (1,3) & & \\
& & (-1,2) & (0,2) & (1,2) & & \\
(-3,1) & (-2,1) & (-1,1) & (0,1) & (1,1) & (2,1) & (3,1) \\
(-3,0) & (-2,0) & (-1,0) & (0,0) & (1,0) & (2,0) & (3,0) \\
(-3,-1) & (-2,-1) & (-1,-1) & (0,-1) & (1,-1) & (2,-1) & (3,-1) \\
& & (-1,-2) & (0,-2) & (1,-2) & & \\
& & (-1,-3) & (0,-3) & (1,-3) & & \\
\end{array}
$$

A *situation* is defined to be a set of pegs on the board. For any situation S we define a *value*

$$A(S) = \Sigma p^{k+l}$$

where the Σ sign indicates that we add up all the p^{k+l} for all the coordinates (k, l) of pegs in the set S. (This, you may note, makes A a *function* with domain the set of possible situations and target the field with 4 elements). Thus for the situation (pegs marked in black)

S is the set $\{(-2, -1), (-1, 0), (0, 0), (0, -1), (2, 0), (1, 2)\}$ and we have

$$
\begin{aligned}
A(S) &= p^{-2-1}+p^{-1+0}+p^{0+0}+p^{0-1}+p^{2+0}+p^{1+2} \\
&= p^{-3}+p^{-1}+p^{0}+p^{-1}+p^{2}+p^{3} \\
&= 1+q+1+q+q+1 \\
&= 1+q \\
&= p.
\end{aligned}
$$

The function A is defined to have the very nice property: if a legal move changes situation S into situation T, then $A(S) = A(T)$. The value of a situation is unchanged by legal moves, and therefore remains constant throughout the game.

To see this, consider a move to the left. Pegs on (k, l) and $(k-1, l)$ are replaced by a single peg on $(k-2, l)$. The value changes by

$$
\begin{aligned}
& p^{k-2+l}-p^{k-1+l}-p^{k+l} \\
&= p^{k+l}(p^{-2}-p^{-1}-1) \\
&= p^{k+l}(p+p^{2}+1) \\
&= 0, \text{ by virtue of (§).}
\end{aligned}
$$

Similarly for moves to the right, up, or down.

There is another function with this property, defined by

$$
B(S) = \Sigma p^{k-l}
$$

(also summed over $(k, l) \in S$). So to each position we can assign a pair

$$
\big(A(S), B(S)\big)
$$

of elements of the field. There are 16 such pairs, and they all occur for suitable positions S. They separate the positions into 16 sets, in such a way that a series of moves does not change which set the positions belong to.

The initial position of the game has $A(S) = B(S) = 1$. So any position that can arise during the game also has $A(S) = B(S) = 1$. For a single peg on (k, l) we get

$$
\begin{aligned}
A(S) &= p^{k+l} \\
B(S) &= p^{k-l}
\end{aligned}
$$

so that we must have

$$
p^{k+l} = p^{k-l} = 1
$$

for any legal final position. The powers of p equal to 1 are p^{-6}, $p^{-3}, p^{0}, p^{3}, \ldots$, and in general p^{3n}. So $k+l$ and $k-l$ are multiples of 3, from which it follows that k and l are multiples of 3. So the only positions we can reach with only one peg left are $(-3, 0)$, $(0, 3)$, $(3, 0)$, $(0, -3)$, and $(0, 0)$.

This does not show that these positions are possible. But it does eliminate plenty that aren't. And in fact all of them *are* possible.

The same analysis works no matter what shape the board is, as long as the holes are arranged in rows and columns. And you can also consider three-dimensional boards in a similar spirit.

Chapter 7　　Symmetry: The Group Concept

*'And now we can solve the problem
without any mathematics at all: just
group theory'* – A Cambridge professor

Many kinds of symmetry occur in nature, and have been recognized from early times. The human figure is approximately symmetrical about a vertical line (more properly, a vertical plane), which is one of the reasons why mirrors seem to invert right and left. This kind of symmetry is known as *bilateral* symmetry.

The Isle of Man's symbol of three running legs, or the swastika, possesses rotational symmetry (Figure 47).

Figure 47

A shape may be symmetrical about several lines at once, or combine bilateral and rotational symmetry. A square is bilaterally symmetric about its diagonals and about lines through the centre parallel to a side: it also can be rotated through 90°.

An entirely different sort of symmetry is exhibited by wallpaper patterns, where the whole pattern can be displaced in various directions without looking any different.

The observation that an object is symmetrical can be of great mathematical power. The discussion of isosceles triangles in Chapter 2 boils down to the assertion that they are bilaterally symmetric. In mathematical physics, laws like the conservation of energy follow from certain (postulated) symmetries of the universe. Such a fundamental property as symmetry should be susceptible of mathematical analysis, and indeed is. The first step is to produce a working definition of symmetry, to make sure we are all talking about the same thing. Otherwise we might confuse 'symmetrical' with 'beautiful' or 'complicated'.

The essence of symmetry is the way shapes can be moved around and still look the same. Individual points, however, need not stay in the same place. If we rotate a square *ABCD* about its centre through a right angle, as in Figure 48, then corner *A* moves to *B*, *B* to *C*, *C* to *D*, and *D* to *A*.

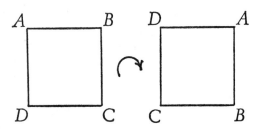

Figure 48

The important thing, then, is not the position of the points, but the operation of moving them. 'Turn through a right angle' describes a symmetry of the square, and so does 'reflect about a vertical line'. Now these are what we earlier called rigid motions, and can be described as certain functions whose range and domain are both equal to the plane $\mathbf{R}^2$.

Thus for any subset *S* of $\mathbf{R}^2$ we can define a *symmetry* of *S* to be a rigid motion $f : \mathbf{R}^2 \to \mathbf{R}^2$ such that for all points $x \in S$, the result $f(x)$ of applying f is also in *S*. We can express this last condition as $f(S) = S$. In geometrical language, a symmetry of *S* is a rigid motion of the plane which leaves *S* in the same place, although it is allowed to move the individual points of *S*.

We need not restrict ourselves to the plane, and three-dimensional space is just as good.

For the running legs there is an obvious symmetry: rotate through 120° about the centre in (say) a clockwise direction. Let us call the corresponding function (or rigid motion) *w*. Another symmetry, say *v*, rotates through 240°. At first sight these are the only possible symmetries, but as always we must watch out for trivial cases too. There is a third symmetry, the identity function.

This leaves every point fixed, and fits our definition, so we must include it. To remind us of its nature, we use the symbol I. The set of symmetries of the running legs is $\{I, w, v\}$.

Rotating through 240° is the same as rotating twice through 120°. In other words, $ww = v$, where the product is defined as in Chapter 5. It simplifies notation if we write w^2 for ww, w^3 for www, and so on; so that $w^2 = v$. In a similar way, $v^2 = w$: if you rotate through 240° twice, the result is the same as if you rotate through 120°, because 360° is a complete rotation and has the same effect as leaving everything fixed. In fact, if we take the 'product' of any two of the symmetries, we obtain a third one. We get a table:

×	I	w	v
I	I	w	v
w	w	v	I
v	v	I	w

(where the entry in row a and column b is ab).

Using this table, we see that $w^3 = I$. This makes sense, because three rotations of 120° take every point back to where it started.

The fact that the product of any two symmetries is also a symmetry is usually expressed as: the set of symmetries is *closed* under the operation of multiplication. If we didn't include I as a symmetry we would lose this property, which would be like having an arithmetic where certain numbers could not be added together to give a number. We *could* do without it, but it's much simpler if we don't.

This set of symmetries, with its multiplication, is an example of a mathematical structure dignified by the title 'group'. We shall define a group later on, but for now all we need is the word. We have found the *symmetry group* of the running legs.

Every shape has a symmetry group. The human figure has two symmetries: the identity, and reflection r about a vertical line. The multiplication table is

×	I	r
I	I	r
r	r	I

and again the set of symmetries is closed under multiplication. Let's do a more complicated example. The equilateral triangle (Figure 49) has six symmetry operations.

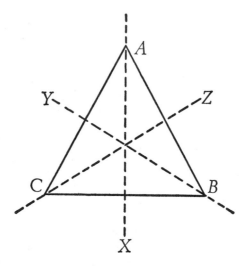

Figure 49

There is the identity *I*, and clockwise rotations *w* and *v* through 120° and 240°. There are also reflections *x*, *y*, *z* in the lines *X*, *Y*, *Z*. (We think of these lines as staying fixed when the triangle moves.) The reflections do not arise when we consider the running legs because they make the feet point the wrong way.

The set $K = \{I, w, v, x, y, z\}$ of symmetries is closed under multiplication (as functions), and we get the table below:

×	I	w	v	x	y	z
I	*I*	*w*	*v*	*x*	*y*	*z*
w	*w*	*v*	*I*	*z*	*x*	*y*
v	*v*	*I*	*w*	*y*	*z*	*x*
x	*x*	*y*	*z*	*I*	*w*	*v*
y	*y*	*z*	*x*	*v*	*I*	*w*
z	*z*	*x*	*y*	*w*	*v*	*I*

For instance, we work out wx as follows: wx means 'do x and then do w'. The triangle

$$A$$
$$C\ \ B$$

moves under x to

$$A$$
$$B\ \ C$$

and then, under w, moves to

$$B$$
$$C\ \ A$$

which gives the same effect as z. So $wx = z$.

Here's another one: yz. Under z the triangle moves to

$$B$$
$$C\ \ A$$

and under y this goes to

$$C$$
$$B\ \ A$$

(remember, the lines X, Y, Z stay fixed). But this is what w does. So $yz = w$.

You should now be able to check that the table is correct. For vividness, make a cardboard triangle and mark its corners; draw round it on a sheet of paper and mark X, Y, Z *on the paper*.

In general, to find the symmetry group of a figure we must

    (i) find all the symmetries,
    (ii) work out the multiplications.

In every case you will find that the set is closed under multiplication. This is no accident. If f and g are rigid motions, so is fg. If f and g both leave the set S fixed, so does fg, because $fg(S) = f(g(S)) = f(S) = S$. If doing f or g doesn't change the shape, neither does doing them both.

The same ideas apply to solid figures. The cube has 24 rotational symmetries, 48 including reflections. We can move any vertex to any other and rotate the edges leading to that vertex in three ways. The dodecahedron has 60 rotational symmetries, 120 including reflections. Naturally we do not bother to work out a multiplication table! There are other ways of expressing the relations between symmetries other than listing all possible products, although we shall not delve into them here.[1]

The Group Concept

The concept of a 'group' is abstracted from these, and other, examples; in the same sort of way that the concept 'ring' was abstracted from arithmetic. Rather than beat around the bush I'll give the definition first, and discuss it after we've seen it.

A *group* consists of:

(1) A set G.

(2) An *operation* '*' which assigns to any elements x and y of G an element $x*y$ *which also belongs to G.*

This operation is required to satisfy three laws:

(3) The operation is associative: for any x, y, $z \in G$ we have
$$x*(y*x) = (x*y)*z.$$

(4) There is an identity element $I \in G$ such that
$$I*x = x = x*I$$
for any $x \in G$.

(5) There are inverses: for any $x \in G$ there exists $x' \in G$ such that
$$x*x' = I = x'*x.$$

Groups can arise in many quite distinct situations. Here are some examples:

(i) G is the set of symmetries of the running legs, and * is multiplication of symmetries, as in the table on p. 97. We have to check properties 1–5. Now (1) is obviously true: we have just defined G to be a set. (2) works because of closure. (3) is true *because it is true for functions*. (4) is true, and we have jumped the gun by using I to denote the identity function. Finally, (5) is true: we can take $I' = I$, $v' = w$, $w' = v$.

(ii) Let G be the set of integers: $G = \mathbf{Z}$. Then (1) holds. Let * be the operation $+$ of addition. Then (2) holds because if a and b are integers then $a+b$ is an integer. Condition (3) is law (1) of arithmetic (p. 76), (4) is law (3) (with 0 playing the part of *I*), (5) is law (4).

(iii) Let $G = \mathbf{R}$, the reals, and * $= +$. Argue as in (ii).

(iv) Let G be the set of non-zero rationals, and let * be the operation of multiplication. G is a set, and the product of two non-zero rationals is a non-zero rational. This deals with (1) and (2). Condition (3) is law (5) of p. 77, which holds for the rationals, and (4) is law (7) (with 1 playing the part of *I*). The rationals form a field, so condition (5) is law (9), which holds in any field.

(v) Let S be any subset of the plane, G its set of symmetries, * multiplication of functions. Then, as in example (i), we have a group.

It should be emphasized that the failure of any of the five conditions means that we do *not* have a group.

If we took G to be the set of integers between -10 and 10, and * to be addition, then (2) is violated: $6+6$ is *not* an element of G.

The set of integers greater than 1, under the operation of addition, has no element satisfying (4).

The set of integers, under the operation of subtraction, violates condition (3) because subtraction is not associative:

$$(2-3)-5 = -6 \neq 4 = 2-(3-5).$$

The set of all rationals, under multiplication, is *not* a group. The only element we can find for I is 1, and then we cannot find an element $0'$ such that $0'0 = 1$; because for any rational r we have $r \times 0 = 0$, so $0'0 = 0$, not 1.

So none of these define groups.

Let me say a few words about the operation *. Given any pair of elements (x, y) where $x, y \in G$ we obtain a unique element $x*y$ of G. This means that * defines a function whose domain is the set $G \times G$ of pairs (x, y), and whose target (in fact range) is G. An operation may be defined as a function

$$*: G \times G \to G$$

as long as we agree that $x*y$ is shorthand for $*(x, y)$. Once we do things this way, condition (ii) is automatic, and may be omitted –

except that we have to check, in any particular case, that * really is a function from $G \times G$ into G.

Once we understand the ideas, we can simplify our notation. Instead of $x*y$ we can write xy (we must remember that this need not be ordinary multiplication) and it then becomes natural to write $x' = x^{-1}$. If you use this notation, but work in the group of integers under addition, then xy means $x+y$ and x^{-1} means $-x$. It is important not to get confused!

Subgroups

If, from the six symmetries of the triangle, we select the three I, w, and v, we find that these form a smaller group inside the big one. You can check this from the multiplication table, or geometrically: these are the symmetries which don't turn the triangle over, and if two such symmetries are multiplied together the resulting symmetry also does not turn the triangle over.

This smaller group within the larger is an example of a *subgroup*. If G is a group with operation *, then a subset H of G is a subgroup if H forms a group under the operation *.

Not every subset H is a subgroup. If we tried $H = \{x, y, z\}$ we would not get a group, because $xy = w$ which isn't in H. If h and k lie in H then we must have

(i) $h*k \in H$
(ii) $h^{-1} \in H$

from which it follows, for non-empty H, that

(iii) $I = h*h^{-1} \in H$.

Conversely, conditions (i) and (ii) are sufficient to ensure that a non-empty subset H is a subgroup, because the associative law is bound to hold in H if it does in G.

Subgroups are extraordinarily common. The group of integers, under addition, has subgroups comprising all even integers, all multiples of 3, all multiples of 4, all multiples of 5, ..., . Every group G is a subgroup of itself, as is the single element set $\{I\}$, which has the trivial multiplication table

$$
\begin{array}{c|c}
\times & I \\
\hline
I & I
\end{array}
$$

The symmetry group of the equilateral triangle has in all six different subgroups:

$$\{I, w, v, x, y, z\}$$
$$\{I, w, v\}$$
$$\{I, x\}$$
$$\{I, y\}$$
$$\{I, z\}$$
$$\{I\}.$$

The number of elements in a group (if finite) is called the *order* of the group. We have just found a group of order 6, having subgroups of orders 1, 2, 3, and 6. It does not take a genius to see that these numbers all *divide* 6. After looking at a few more examples, one is tempted to guess that the order of a subgroup always divides the order of the group.

This guess would be correct: the theorem was proved by Lagrange some time *before* the abstract concept of a group was defined!

Let's put $K = \{I, w, v, x, y, z\}$ and consider the subgroup $J = \{I, x\}$. For any element $a \in K$ we form the *coset $J*a$*, defined to be

$$\{I*a, x*a\}.$$

We form this by multiplying each element of J by a, and collecting the resulting elements into a single set. We can compute what these are:

$$
\begin{aligned}
J*I &= \{I, x\} & J*x &= \{I, x\} \\
J*v &= \{v, z\} & J*z &= \{v, z\} \\
J*w &= \{w, y\} & J*y &= \{w, y\}.
\end{aligned}
$$

We notice several things:

(i) There are only 3 distinct cosets.
(ii) One of them is J itself.
(iii) No distinct cosets have any element in common.

(iv) Every element of K lies in some coset.

(v) Each coset has the same number of elements.

From (ii) and (v) each coset has 2 elements. From (iii) and (i) all the cosets, taken together, have $2.3 = 6$ elements. From (iv), K has 6 elements. This not only explains why the order of J divides that of K; it says that the result of doing the division will be the number of cosets.

The proof of Lagrange's theorem follows the same lines. One shows that (i)–(v) are true for any group K and any subgroup J (except that in (i) we get an unknown number c of cosets). If J has order j and K has order k, it follows that $k = jc$. So j divides k. Properties (ii)–(v) follow from the group axioms, after a little preparatory work.

This is a remarkable theorem. From the vague-sounding (though actually hyperprecise) abstract ideas of group and subgroup we extract a concrete *numerical* relation. If I gave you a group of order 615 you would know, *without* any information about the multiplication table, that its subgroups cannot have any orders other than 1, 3, 5, 15, 41, 123, 205, and 615.

One might ask whether all these must occur. The group of rotations of the dodecahedron has order 60, but has no subgroup of order 15, even though 15 divides 60. The best that can be said in general is *Sylow's theorem*: if h is a power of a prime and divides the order of a group G, then G has a subgroup of order h. So our group of order 60 certainly has subgroups of orders 2, 3, 4, and 5. Any group of order 615 has subgroups of orders 3, 5, and 41.

Isomorphism

There are other ways of producing groups with 6 elements. If we take the set $S = \{a, b, c\}$ there are 6 *bijections* $S \to S$, namely the functions p, q, r, s, t, u given by

	p	q	r	s	t	u
a	a	a	b	b	c	c
b	b	c	a	c	a	b
c	c	b	c	a	b	a

where, for example, the value $s(b)$ of the functions s at b is the entry in row b, column s: namely, c.

Bijections between a set and itself are known as *permutations* of the set.

Under multiplication of functions these 6 bijections form a group, with multiplication table

×	p	q	r	s	t	u
p	p	q	r	s	t	u
q	q	p	t	u	r	s
r	r	s	p	q	u	t
s	s	r	u	t	p	q
t	t	u	q	p	s	r
u	u	t	s	r	q	$p.$

For instance, to find rs we have

$$rs(a) = r\big(s(a)\big) = r(b) = a$$
$$rs(b) = r\big(s(b)\big) = r(c) = c$$
$$rs(c) = r\big(s(c)\big) = r(a) = b$$

which makes rs have the same effect as q. So $rs = q$.

This is not the *same* as the symmetry group of the equilateral triangle, because its elements are different. But there is a strong resemblance between the two groups, in addition to them both having order 6.

Every symmetry of the triangle rearranges the vertices A, B, C, in the following way:

	I	w	v	x	y	z
A	A	B	C	A	C	B
B	B	C	A	C	B	A
C	C	A	B	B	A	C

This suggests we pair off rigid motions with permutations, by changing capital A, B, C to small a, b, c:

$$
\begin{array}{cccccc}
p & q & r & s & t & u \\
\updownarrow & \updownarrow & \updownarrow & \updownarrow & \updownarrow & \updownarrow \\
I & x & z & w & v & y.
\end{array}
$$

If we rewrite the two multiplication tables so that elements paired off here occur in corresponding positions along the top and down the sides, and then fill in the products from the original tables, we get:

×	I	w	v	x	y	z
I	I	w	v	x	y	z
w	w	v	I	z	x	y
v	v	I	w	y	z	x
x	x	y	z	I	w	v
y	y	z	x	v	I	w
z	z	x	y	w	v	I

×	p	s	t	q	u	r
p	p	s	t	q	u	r
s	s	t	p	r	q	u
t	t	p	s	u	r	q
q	q	u	r	p	s	t
u	u	r	q	t	p	s
r	r	q	u	s	t	p.

Not only do the elements occur in corresponding positions along the top and down the sides: they also occur in corresponding positions in the main body of the tables. Thus I and p occur in the positions

and x, q occupy

One should not be too surprised at this, because the way in which the permutations multiply is very closely connected with the way the symmetries multiply. It shows that two groups can have the same structure without being identical. The difference between them lies in the *names* of the elements.

To make this idea sufficiently precise to be useful, we consider the function f such that $f(I) = p, f(x) = q, \ldots$, which gives the correspondence between elements of the two groups. The domain of *f* is the first group, its range is the second. Take two elements *a* and *β* of the first group. In row *a* and column *β* we get the element *a*β*. In the corresponding row *f(a)* and column *f(β)* of the second table, we should get *f(a)*f(β)*. But this, we observed, is the element corresponding to *a*β*, which is *f(a*β)*. So the observation that 'corresponding elements occur in corresponding places' means that

$$f(a*\beta) = f(a)*f(\beta) \tag{†}$$

for all *a* and *β* in the first group.

The advantage of using (†) is that it does not depend on geometrical properties of multiplication tables. Given any two groups *G* and *H* we say that they are *isomorphic* if there is a bijection *f*: *G* → *H* such that (†) holds for all *a*, *β* ∈ *G*. Isomorphic groups have the same abstract structure, and differ only in their elements. Since it is the way the elements multiply that contain the essential structure of the group, for most purposes isomorphic groups may be thought of as being the same.

The first of our two groups above, we found, has 6 subgroups. This immediately implies that the second, which is isomorphic, has 6 subgroups too: for instance, the subgroup {*I, w, v*} of the first gives rise to a subgroup {*p, s, t*} of the second.

Groups of equal order need not be isomorphic. There is another group of order 6, which we can get by taking the integers mod 6 under the operation of addition. Its 'multiplication' table is the addition table mod 6, namely

+	0	1	2	3	4	5
0	0	1	2	3	4	5
1	1	2	3	4	5	0
2	2	3	4	5	0	1
3	3	4	5	0	1	2
4	4	5	0	1	2	3
5	5	0	1	2	3	4

Call this group *M*. Is *M* isomorphic to the symmetry group *K* of the equilateral triangle?

One way to decide would be to try all possible bijections from

M to K and see if equation (†) holds. If we tried defining f by $f(0) = I, f(1) = w, f(2) = v, f(3) = x, f(4) = y, f(5) = z$, then we would have

$$f(1+2) = f(3) = x$$
$$f(1)*f(2) = wv = I$$

which means we have the wrong function. There are only 720 bijections to try. It could be done that way.

Instead, we could try to find properties of M which do not depend on the names of the elements. One such property, we saw, was how many subgroups it has. If you work them out, you will find that the subgroups of M are $\{0\}$, $\{0, 2, 4\}$, $\{0, 3\}$, and M. So M has only 4 subgroups, and cannot therefore be isomorphic to K, which has 6.

This is better than trying 720 functions: But there is an easier way. Addition mod 6 satisfies the commutative law (law (2) of p. 76), $a+\beta = \beta+a$. Suppose we had an isomorphism $f: M \to K$. Then

$$f(a+\beta) = f(\beta+a)$$

so that from (†)

$$f(a)f(\beta) = f(\beta)f(a).$$

In other words, K also would satisfy the commutative law, this time for multiplication. But $vx = y$, $xv = z$, so it doesn't. So no isomorphism can exist between group M and the symmetry group K.

Isomorphisms are a great simplifying device. It is important in mathematics to be able to recognize when two apparently distinct problems are basically the same. If isomorphic structures occur in the two problems, this may give a hint of connections between them.

In our example above we found the isomorphism by exploiting a known connection between the symmetries of the triangle and the permutations of 3 elements. Sometimes it happens the other way round: you observe an isomorphism and ask why it happens. There is a certain group of permutations on a set with 5 elements, associated with the general equation of degree 5. (The elements permuted are the 5 roots of the equation.) This group has 60 elements. The group of rotations of the dodecahedron also has 60 elements. It can be shown that the two groups are isomorphic.

Starting from this coincidence Felix Klein[2] discovered a deep interconnection between three theories:

> The quintic equation,
> rotation groups,
> the theory of complex functions.

Among other things, this explained a previously observed fact: the quintic equation can be solved using a certain special kind of complex function – known as elliptic functions. Before Klein's synthesis, this could be proved only by formless calculations. It, too, seemed like a coincidence. Klein found out *why* it happened.

Classifying Patterns

Group theory will crop up whenever symmetries exist. It allows us to describe symmetries according to the underlying group. By dodecahedral symmetry, for example, we mean 'having a symmetry group isomorphic to that of the dodecahedron'.

Not only this. It allows us to classify symmetries. In certain situations we can say: these symmetries, and only these, are possible.

Wallpaper patterns, abstractly, are symmetric configurations in the plane. The symmetry group of a wallpaper pattern consists of certain rigid motions, and is a subgroup of the group *G* of *all* rigid motions of the plane. To go further, we must say more carefully what we mean by wallpaper patterns: they must extend arbitrarily far, and they must be *discrete* in the sense that they 'come in lumps' instead of merging continuously into each other. (There is a precise but technical mathematical description.) One can then classify the suitable groups of rigid motions, with the conclusion that there are precisely 17 wallpaper patterns (9 of which are 'friezes' rather than genuine wallpapers). These are illustrated in Figures 50–52.

When you look at a book of wallpaper samples – hundreds and hundreds of different patterns – you would not imagine that any useful classifications could be made. There are so many of them. But if you forget about the colours, the size, the quality of the

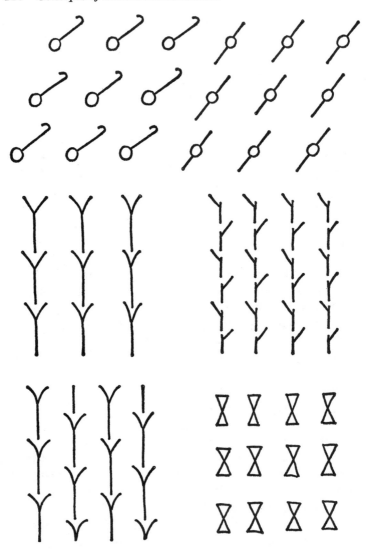

Figure 50

paper (all of which are relevant to the practical side of wall-papering!) and concentrate on the basic structure, this fact emerges: there are only 17 basically different kinds. It is alleged that all 17 occur in the works of Arabian potters. It would be an

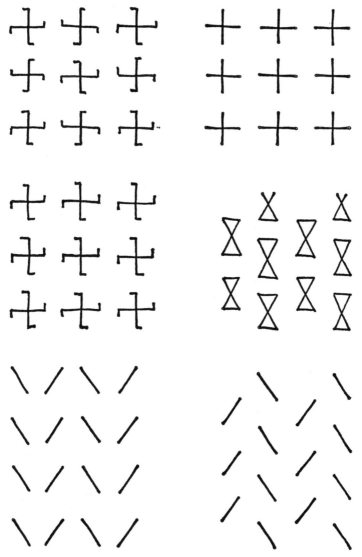

Figure 51

interesting exercise to get hold of a wallpaper pattern book and
see whether today's designers are similarly exhaustive. Probably
not.

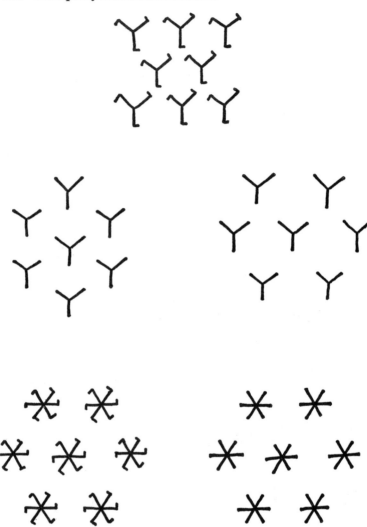

Figure 52

The analogous problem in three dimensions – the classification of the 230 possible symmetry groups[3] – is important in crystallography: from this one can make deductions about the molecular structure of crystals.

Chapter 8　Axiomatics

> *'Only an elephant or a whale gives birth*
> *to a creature whose weight is 70*
> *kilograms or more. The President's*
> *weight is 75 kilograms. Therefore the*
> *President's mother was either an*
> *elephant or a whale'* – Stefan Themerson

Mathematics operates on many levels. A child learns how to work out problems involving a particular number or numbers. Later, he deals with properties common to *all* numbers: in a sense the object with which he works has changed from a *number* to the ring $\mathbf{Z}$ of all numbers. Then, in ring theory, instead of studying one particular ring, he studies whole *classes* of rings. A whole area of mathematics becomes a single object, and that object is but one of many in another area. And so it goes on.

Our line of thought in this chapter will take us one stage further: the objects of our thought will be complete theories – ring theory, field theory, group theory, geometry.

There are many similarities in the ways we have defined the concepts 'group', 'ring', 'field'. We introduce certain basic terms. These terms are never defined. Instead we list a number of laws that they are required to satisfy. These laws are the *axioms*; the whole set-up is an *axiomatic system*.

You are not asked to 'believe' the axioms. Indeed, it is futile to question them, as well as irrelevant; because they do not correspond to reality. Whenever you are confronted by an axiomatic system, somebody else is telling you what properties *he* means the system to have. Axioms are like the rules of a game. If you changed them, you would no longer be playing the *same* game.

Starting from a system of axioms, one then makes certain logical deductions. All of these take (implicitly or explicitly) the form: *if* the axioms hold, *then* something else does. The 'truth' of the axioms is not in dispute. The fact that the Roman empire *did* collapse is not relevant to a discussion of what *might* have happened had it *not* collapsed. What one can dispute is the validity of the deductions.

One can also dispute the applicability of the axioms in any particular case. Whether or not the real world works the way the axioms say is a pertinent question *when one tries to apply the theory to the real world*, but it is not a question that is part of the theory. It needs to be answered by experiment. In the same way, in order to apply group theory to a branch of mathematics, one must check that the relevant objects *are* groups. If they are not, we cannot apply the theory. But that makes no impression on the theory. 'Are the group axioms true?' is a nonsense question. Axioms are not true in any absolute sense; but they may be true *of* something.

The power of the axiomatic method is that it derives a large body of theory from a small number of assumptions. If anything satisfies the assumptions, it is bound to satisfy all the conclusions derived from them. We can apply the whole power of the theory at the expense of checking just a few properties: we don't have to go through all the work over and over again for each application.

The concept of an axiom system as something divorced from reality is relatively recent. It would seem that the ancient Greeks, when they laid down axioms for geometry, thought that they were talking of genuine physical truths, albeit of an idealized nature. Certainly the common definition of 'axiom' was 'self-evident truth' – and my dictionary still maintains that this is the case. But the word as used in mathematics has taken on a different meaning. Turn to the group axioms. Are they self-evident?

Euclid's Axioms

Euclid listed a number of axioms for geometry, of which the most important are:

(1) Any two points lie on a straight line.

(2) Two lines meet in at most one point.

(3) Any finite line segment may be produced as far as you wish.

(4) It is possible to describe a circle with any centre and any radius.

(5) All right angles are equal.

(6) Given any line, and any point not on the line, then there exists *exactly one* line parallel to the first line and passing through the given point.

(These are not given in exactly the form Euclid gave.)

It was for a long time regarded as a blemish that axiom (6) appeared to be far from self-evident. Many attempts were made to prove it from the other axioms: all fallacious.

We shall see later that axiom (6) cannot be proved in this way. A more meaningful question is: is it true of the real world? This is not a mathematical question. To answer it we should perform an experiment. However, imagine the ancient Greeks performing such an experiment. They draw two 'parallel' lines – say lines of longitude – through Rome and Athens: these meet at the South Pole. The parallel-axiom is false for the geometry of the surface of the earth.

This is cheating, really: we know that the earth is round, and Euclidean geometry applies to planes, not spheres. Actually, if you think about it, we know the earth to be round precisely because it does not behave as Euclidean geometry says; so if Euclid was wrong then perhaps the earth need not be round after all.

A fairer experiment would be to use laser beams, or some such, as straight lines. You point your lasers out into intergalactic space, as nearly parallel as you can get them, and then try to find out if they meet or not. This, unfortunately, is not an experiment that could be carried out in practice. (And if some cosmologists are correct, it would not verify Euclidean geometry even if we could carry it out.)

It looks as if Euclid had a far better idea of what he was up to than many who later criticized him. He must have suspected that the parallel-axiom could not be proved, and that is why he stated it explicitly.

Consistency

When you first start working out an axiomatic theory, all you have to go on is the axioms (as far as logical deduction is concerned: psychologically you will have some intuitive ideas as to how the theory should develop). You use these to prove some theorems; then use these theorems to prove others. The axioms become the source of an outward-spreading wave of theorems, all ultimately dependent upon them.

All is well as long as you cannot prove two contradictory theorems in this way. But if you *can* prove two contradictory theorems, then the whole theory becomes useless. For it is then possible to prove *anything*.

The analyst G. H. Hardy once made this remark at dinner, and was asked by a sceptic to justify it: 'Given that $2+2 = 5$, prove that McTaggart is the Pope.' Hardy thought briefly, and replied, 'We also know that $2+2 = 4$, so that $5 = 4$. Subtracting 3 we get $2 = 1$. McTaggart and the Pope are two, hence McTaggart and the Pope are one.'

To proceed more generally we must first recall the method of *proof by contradiction* (or *reductio ad absurdum*). We wish to prove a statement *p*. We begin by assuming that *p* is false: on the basis of this we deduce two contradictory statements. This is absurd, so our assumption of the falsity of *p* must be wrong. Therefore *p* is true. The validity of this method is built in to current mathematical logic. We used it in Chapter 6 to prove $\sqrt[3]{2}$ irrational. But now suppose we have an axiomatic system from which we can deduce two contradictory theorems *r* and *s*. Perhaps *r* is 'butter is cheap' and *s* is 'butter is *not* cheap'. Then we can use *these* to provide the contradiction in the above proof, *whatever p* may be. The theorems *r* and *s* can be deduced from *p*, because they can be deduced from the axioms. We don't actually have to *use p* in the deduction.

For instance, to prove 'the country is going to the dogs' we assume the opposite: it is *not* going to the dogs. Now we deduce both 'butter is cheap' and 'butter is not cheap', which contradict each other. Our assumption must be wrong, so the country *is* going to the dogs.

The same argument will also prove that it is *not* going to the dogs: start by assuming it is and proceed as before.

This is total disaster. One could, perhaps, put up with an oracle which occasionally answered both 'yes' and 'no' to the same question. But what use is an oracle which *always* answers 'yes'?

A system of axioms which does not contradict itself is said to be *consistent*. Consistency is a prime requisite for any axiomatic theory. Its importance was first emphasized by David Hilbert, the founder of the modern theory of axiomatics.

It is not always obvious that an inconsistent theory *is* inconsistent. The problem can be a very delicate one. The axioms for a field are consistent. But if we modify law 9 to read 'every element [rather than every non-zero element] has a multiplicative inverse', then the system becomes inconsistent. Because if 0 had an inverse 0^{-1} we would have

$$(0.0).0^{-1} = 0.0^{-1} = 1$$
$$0.(0.0^{-1}) = 0.1 = 0$$

which contradicts the associative law, because axiom (10) of p. 80 says that $0 \neq 1$. (This is why 'thou shalt not divide by zero'. It messes up the laws of arithmetic.)

So here we have a very small modification which changes a consistent system into an inconsistent one. And unless you know where to look, the second system is not obviously inconsistent. However innocuous a given set of axioms may appear, the question of its consistency will still arise.

Models

Hilbert also gave two other requirements for an axiomatic system; *completeness* and *independence*.

To say what 'completeness' is we need the idea of a *proof* in an axiomatic system. If p is some statement in the system, a proof of p consists of a sequence of statements, each of which is either an axiom or a logical consequence of certain *preceding* statements in the list, such that the last statement in the list is p. The proof of the statement

$$(x+y)^2 = x^2+2xy+y^2$$

in Chapter 6 is an example. A system is complete if, for *every* statement p, we can find a proof of p, or a proof of not-p. In other words, we have got enough axioms to *prove* the truth or falsity of any conceivable statement of the system.

In a complete system there is no significant way in which we can add extra axioms: either they will follow from those we already have, and so be redundant; or they will contradict those we already have, and so be pointless.

A set of axioms is *independent* if no axiom can be deduced from the others.

It is always difficult to prove an axiom system complete, if indeed it is, because one needs to consider all possible proofs. But there are simple methods for proving independence (in suitable circumstances), and occasionally consistency. These revolve around the idea of a *model*.

A model of an axiom system is some object in which, with a suitable interpretation, the axioms are true. Any group is a model for the group axioms: the abstract operation * of the axioms is interpreted as some definite operation in the special group under consideration. It might be addition, or multiplication of functions. Similarly any ring is a model for the ring axioms, and any field a model for the field axioms. Coordinate geometry provides an algebraic model for the axioms of Euclidean geometry, if we interpret 'point' to mean 'pair (x, y) of real numbers' and 'line', 'circle', etc. in the usual way.

If you can exhibit a model for a system of axioms, then they must be consistent. Any one of the group multiplication tables defines a model: we could take the most trivial possible example

$$
\begin{array}{c|c}
\times & I \\
\hline
I & I
\end{array} .
$$

If the axioms were inconsistent, then *any* theorem could be proved. You could prove that every group has 129 elements. But since the axioms hold in the model, so do all consequences of them. So the model must have 129 elements. But we can see that it doesn't. Therefore there are no inconsistencies.

Or, to argue slightly differently, any contradiction in theorems deduced from the axioms will show up in the model. We will be able to prove that the model has a certain property, and also that it does not. This can't happen, because the model either has the property or not: it can't have both.

Models are particularly useful for proofs of independence. Suppose we wanted to prove that the associative law for group multiplication is independent of the other axioms for groups. All we need do is find a model in which the associative law does not hold, but the other axioms do. Any deduction of the associative law from the other axioms would lead to a proof that the multiplication in the model was associative: but we have chosen our model so that it isn't.

We'll define a model using a multiplication table. We want to satisfy axioms (1), (2), (4) and (5) of p. 100, but not (3).

Axiom (1) says we need a set G. To make life easy, we take a small set; but to give room to manoeuvre we won't make it *too* small. Try $G = \{a, b, c\}$.

Axiom (4) asks for an identity element. If we make a the identity, then part of the table is determined:

$\times$	a	b	c
a	a	b	c
b	b		
c	c		

Next, look at axiom (5), which asks for inverses. Our identity element a already has an inverse, because $a^2 = a$. If we arranged to have $bc = cb = a$ this would provide inverses for b and c. So now our table looks like

$\times$	a	b	c
a	a	b	c
b	b		a
c	c	a	

and axioms (1), (4), and (5) hold.

Now axiom (2) says that we must define the product of any two elements, and that product must be in G. To satisfy this, all we need do is fill in the rest of the table with as, bs, and cs. It doesn't matter how we do it. But since we want (iii) to be false, we must avoid choices which make (iii) hold. It would not do to fill the two remaining spaces with c and b, because the table would look like that for the symmetries of the running legs, which *do* satisfy the associative law. So we try

×	a	b	c
a	a	b	c
b	b	b	a
c	c	a	b

After a few trials we discover that for this table

$$(cc)b = bb = b$$
$$c(cb) = ca = c$$

so the associative law fails.

This completes the construction of the model.

The construction of models is an art, rather than a science. It requires experience, taste, and a dash of de Bono's 'lateral thinking'. The best way to learn how to do it is to try.

We shall return to the questions of completeness and consistency in Chapter 20. Our immediate object is to apply the method of models to the problem of Euclid's parallel-axiom.

Euclid Vindicated

We can phrase the problem thus: is Euclid's parallel-axiom independent of his other axioms? Having stated it in these terms, we are half way to a solution: the biggest difficulty is realizing that independence might be the case. The two alternatives: it is provable from the other axioms, it is disprovable from the other axioms, do not exhaust the possibilities.

For our answer we must make one assumption: that the axioms for Euclidean geometry are consistent. This is because we are going to use Euclidean geometry as the raw materials for the model. And if it was inconsistent, the question of independence would not be our main cause of worry.

Construction of models, I said, was an art. On this occasion the art is conjuring: I can do no better than wave the magic wand and extract the rabbit from the hat.

Draw a circle Γ in the plane. Our model will be that part of

Euclidean geometry which happens *inside* Γ. To keep the record straight we'll use italic type for the interpretations, in the model, of standard Euclidean concepts. Let us define

point	= point of the plane inside Γ
line	= that part of a line in the plane which lies inside Γ
circle	= that part of a circle in the plane which lies inside Γ
right-angle	= ordinary right-angle, inside Γ.

(These are illustrated in Figure 53.)

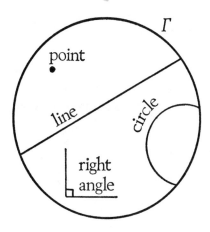

Figure 53

Now we check the axioms (as on p. 114).

(1) Any two *points* lie on a straight *line*. This is true. If we take two *points* in Γ they are also points in the plane. We join them by a line in the plane, and then chop off the bits outside Γ to get a *line* (Figure 54).

(2) Any two *lines* meet in at most one *point*. This is true. The two *lines* are parts of two lines, which meet in at most one point, so certainly in at most one *point* (Figure 55).

(3) Any finite *line* segment may be produced as far as you wish. This is more controversial. At first sight it seems to go wrong, because as soon as a *line* goes outside Γ it ceases to be

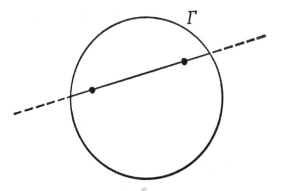

Figure 54

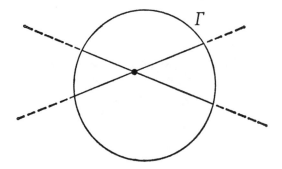

Figure 55

a *line*. But even Euclid did not mean you to produce lines *outside the plane* – off the edge, so to speak. The construction must be confined to the region under discussion. So we need to consider the concept *produce* rather than produce. The force of axiom (3) is that if you have a line segment with ends, you can extend it beyond the ends. This is also true in Γ, provided that when we say 'inside Γ' we do not include anything *on* Γ. Because, as in Figure 56, we can keep extending the *line* to points 1, 2, 3, 4, 5, . . . , without ever stopping.[1]

The *line* itself has no ends: what ought to be its ends lie on Γ, not inside it, so are not *points*. As far as the model is concerned all *lines* go on for ever.

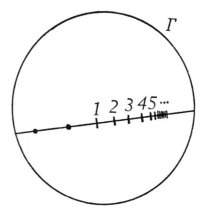

Figure 56

(4) It is possible to describe a *circle* with any centre and any radius. This follows from the same axiom in the plane, as always chopping off anything outside (or on) Γ. Of course *circles* are not always 'circular' (Figure 57).

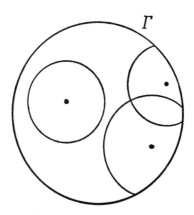

Figure 57

But this does not affect the validity of the axiom.

(5) All *right angles* are equal, again because in the plane all right angles are equal.

Thus the model satisfies axioms (1)–(5). However, axiom (6) is *not* satisfied: Figure 58 shows a *line*, a *point*, and several *lines* parallel to the first *line* through that *point*.

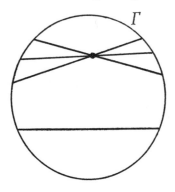

Figure 58

Here, of course, we interpret 'parallel' as 'not meeting'. It does not matter that the *lines*, if produced *outside* Γ, meet; for points outside Γ are not a part of the model.

It now follows that axiom (6) is not provable from axioms (1)–(5). For if it were, then since (1)–(5) are true for the model, then (6), as a logical consequence, must also be true for the model. But it isn't. From a slightly different point of view: any proof of axiom (6) in Euclidean geometry becomes a proof of axiom (6) in the model if we replace all occurrences of 'point' by '*point*', of 'line' by '*line*', and so on. Since (6) is false in the model the supposed *Euclidean* proof cannot exist. The fact that '*points*' are not the same as 'points' does not affect the argument: the difference between them is taken care of by the validity of axioms (1)–(5).

This is what I meant by the section heading 'Euclid Vindicated'. Not that Euclidean geometry is the only one possible. But that Euclid was quite right to put down the parallel-axiom as an *assumption*, *not* provable from his other axioms.

Other Geometries

By making different choices of model you can polish the proof up a bit; in particular making (3) and (4) more convincing. The trick is to redefine length inside Γ to make all lines infinitely long (though then they have to be bent). For more details, consult Sawyer.[2]

Instead of having more than one parallel, we can arrange to have none at all. The model this time is due to Klein: we use small capitals for interpretations in the model.

Construct a sphere Σ in three-dimensional space. The surface of Σ will play the part of the Euclidean plane. Define

> LINE = great circle on Σ (that is, one whose centre coincides with the centre of the sphere);
>
> POINT = pair of diametrically opposite points on Σ.

Now we check the axioms. Axiom (2) is true because any two great circles meet in a pair of diametrically opposite points, and the other axioms (1), (3), (4), (5) hold (with no quibbles about (3)). But because any two great circles meet, there are no parallel LINES at all! (See Figure 59).

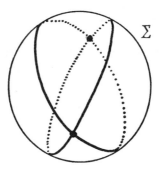

Figure 59

So now we have three different kinds of geometry: Euclidean geometry; *hyperbolic* geometry where there may be many different parallel lines; and *elliptic* geometry, where there are none.

Riemann introduced a more general kind of geometry, which can be elliptic in some parts and hyperbolic in others. The two-dimensional version of this can be thought of as the geometry of a curved surface (Figure 60).

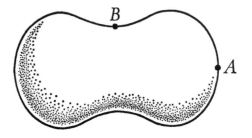

Figure 60

Near *A* the geometry is elliptic, near *B* it is hyperbolic. (This explains the terminology: a section of the surface near *A* is roughly an ellipse, near *B* a hyperbola.)

Riemann's idea goes further than this. There are spaces of three dimensions (or more) whose geometry also varies from place to place. 'Curved' space! According to Einstein, ordinary space–time forms such a geometry. We can either say that the 'curvature' is caused by gravitational attraction of matter, or that matter and gravity are caused by the curvature.

If the geometry of space–time is elliptic in some regions, it may be possible to set out in a straight line and eventually get back where you started. Worse, the same could happen to time: you might get back before you set out. This may sound unlikely. But some astronomers claim that already there is a higher-than-chance proportion of radio stars at diametrically opposite points in the sky. It could be that these pairs of stars are really just *one* star seen from two opposite directions.

Chapter 9
Counting: Finite and Infinite

'Fourteen,' said Pooh. 'Come in.
Fourteen. Or was it fifteen? Bother.
That's muddled me' – A. A. Milne

You cannot teach a child to count by telling him what *numbers* are. Instead, you show him instances of their occurrence: two dogs, two apples, two books, . . . , and he gradually observes that the property of 'two-ness' is common to all these examples. He forms – for himself – the concept 'number'.

Numbers are properties of *sets*. It is the set of apples, or dogs, that has two elements; not any individual apple or dog. We do not count an object: we count sets of objects. When mathematicians began to wonder what numbers really were, they observed this fact. They also realized that it is easier to say when two numbers are the same than to say what they are.

If a child has two cups, each sitting in its own saucer, there comes a stage when he realizes that he must also have two saucers. Playing musical chairs, if there are seven players and six seats, somebody won't have a seat. If a theatre manager sees that each seat in his theatre is occupied by exactly one person, then he knows that the number of people is exactly the same as the number of seats. He doesn't have to know how many seats there are to know this.

This means that the concept 'same number' does not depend on the concept 'number' (despite the vagaries of the English language). In the same way, you can find out if two pieces of string have the same length without ever knowing what that length is, by laying them side by side. Or you can tell if two objects have the same weight, using a beam-balance. In all three cases it is easier to say when two given objects have the property in common than to say in general what that property is. All you need is a way of comparing objects with respect to the (as yet undefined) property.

For length or weight it is not hard to decide the method of comparison. What of 'number'?

Let's go back to the example of people in theatre seats. In order to be certain that the numbers are exactly the same, we need to know:

(i) Each person sits in exactly one seat.

(ii) Each seat contains exactly one person.

If we let S be the set of seats and P the set of people, then for each person $p \in P$ we can define $f(p) \in S$ to be the seat in which he is sitting. And then $f: P \to S$ is a bijection (one-to-one correspondence). Firstly, f fits our definition of a function: its domain is P, its target S. The rule assigning $f(P)$ to p is unambiguous, by (i) above. And f is a surjection, because by (ii) each seat contains a person, and an injection, since by (ii) again it contains only one person.

Quite generally, two sets will have the same number of elements if and only if there is a bijection between them. The situation is illustrated by Figure 61.

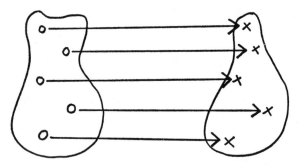

Figure 61

To avoid linguistic problems, let us say that two sets are *equinumerous* if there exists a bijection between them. This carries the same implications as 'same number' but makes it clearer that we don't yet need to know what a number is. In the same way that we can chop up the set **Z** of integers into congruence classes, we can chop up the set of all sets into classes, in such a way that two sets are equinumerous precisely when they lie in the same class. Each class may be specified by giving one of its members. The class containing $\{a, b, c, d, e\}$ will also contain every set equinumerous with $\{a, b, c, d, e\}$, and these are exactly the sets with 5 elements. The situation is illustrated in Figure 62.

Figure 62

In this sense the number 5 is specified by

> (i) giving some set and saying that it has 5 elements;
> (ii) saying that any set equinumerous with the given one also has 5 elements.

In fact, as Frege observed, we have a very curious situation. There is a mysterious and wonderful concept, 'number', which we cannot define. There is a down-to-earth concept, the class to which a given set belongs. Two sets have the same number if and only if they belong to the same class. From this it follows that if we know everything about the classes we know everything about the numbers.

At this point one can adopt two attitudes.

> (A) Whatever these silly classes are, I know perfectly well that they aren't numbers. They just behave like them.
>
> (B) I don't know what numbers are: I just use the word.

These classes behave just like my hypothetical numbers, with the advantage that I know what they are. I might as well say that numbers *are* these classes.

It doesn't matter much which attitude we adopt, as long as we realize that (B) has a good point: we *could* use the classes to define numbers if we wanted to, and it would give a perfectly satisfactory definition.[1] Indeed, when we point out to a child two dogs, two apples, two books, are we not just pointing out the elements of the class associated with '2'?

All we really need to know about numbers is this: to every set is associated something called its number. This has the property that two sets have the same number if and only if they are equinumerous.

We could take the existence of numbers, with the above property, as an axiom. Once we know this, we can recover all of arithmetic. First we define a few numbers:

> 0 is the number of the empty set ∅
> 1 is the number of the set $\{x\}$
> 2 is the number of the set $\{x, y\}$
> 3 is the number of the set $\{x, y, z\}$
> 4 is the number of the set $\{x, y, z, w\}$
> . . .

where of course $x, y, z, w, \ldots$, are chosen to be distinct.

Then we go on to define addition and multiplication. To see how to do this, we go back to primary school. There we added 3

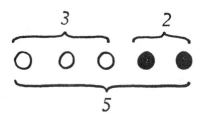

Figure 63

and 2 by taking 3 counters, and then 2 more counters, putting them in a row, and counting the result (Figure 63).

It is important that all the counters should be distinct. If one of our 2 counters was already in the set of 3 counters, we would get the wrong answer!

We can define addition of any two numbers in the same way. Take two numbers m and n. Find sets M and N with these numbers (respectively) and such that they are *disjoint*: they have no elements in common. (In set-theoretic symbols, $M \cap N = \emptyset$.) Form the union $M \cup N$. This has a number: and we define this to be $m+n$ (Figure 64).

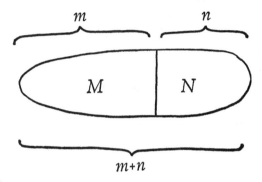

Figure 64

Two aspects of this definition require comment. The first is that we always can find *disjoint* sets M and N. For if they are not disjoint, we can change elements one by one until they are; and the way we change them defines a bijection between the old sets and the new ones, which ensures that the numbers don't change.

The second is that addition should be 'well-defined'. If we took different sets M' and N' with numbers m and n, would we get the same result? If not, the definition would be useless. It's a very poor definition of addition that tells you $2+2 = 4$ if you work it out one way, and $2+2 = 5$ if you work it out the other.

All right: suppose we do pick different sets M' and N', disjoint, with numbers m and n. Then M and M' have the same number, so there is a bijection $f: M \to M'$. Similarly there is a bijection $g: N \to N'$. But we can fit these together to give a bijection $h: M \cup N \to M' \cup N'$, if we set

$$h(x) = \begin{cases} f(x) & \text{if } x \in M \\ g(x) & \text{if } x \in N. \end{cases}$$

(It is necessary to check that this *is* a bijection. It is intuitively obvious from Figure 65.)

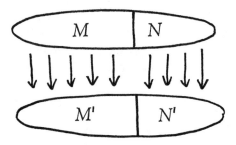

Figure 65

We already have enough information to prove a famous theorem: that $2+2 = 4$.

First we must find disjoint sets M and N, each of which has number 2. From the definition of 2 we may take $M = \{x, y\}$. For N we take the set $\{a, b\}$ where a, b, x, y are all different. There is a bijection $f: M \to N$ such that $f(x) = a, f(y) = b$, so M and N are equinumerous. By the main property of numbers, N has number 2 as well. Next we form $M \cup N = \{x, y, a, b\}$. There is a bijection g from this to the set $\{x, y, z, w\}$ defined by $g(x) = x$, $g(y) = y$, $g(a) = z$, $g(b) = w$. This latter set by definition has number 4, so $M \cup N$ has number 4. By the definition of $+$, $2+2 = 4$.

The basis of the argument is illustrated in Figure 66.

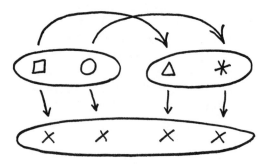

Figure 66

It is slightly easier to define multiplication, because as it turns out we don't need to worry about disjointness. To multiply numbers *m* and *n*, take *M* and *N* with these numbers, and form the Cartesian product $M \times N$ (Chapter 4). Define *mn* to be the number of $M \times N$. That this is the right definition becomes clear when we consider the diagram used in Chapter 4 to illustrate the Cartesian product (Figure 67).

(It is once more necessary to check that a different choice for *M* and *N* gives the same answer, but this is not hard.)

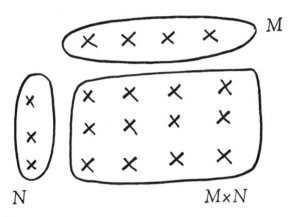

Figure 67

One of the more impressive feats made possible by this line of approach is that we can prove that the various laws of arithmetic hold, at least for positive numbers. (The laws for negative numbers, rationals, or reals can be deduced from these.)[2] As an example, take the distributive law

$$(m+n)p = mp+np.$$

We take sets *M*, *N*, *P* with numbers *m*, *n*, *p*, such that *M* and *N* are disjoint. Then $(m+n)p$ is the number of the set $(M \cup N) \times P$, while $mp+np$ is the number of $(M \times P) \cup (N \times P)$. Now it so happens that these two sets are equal: the first consists of all ordered pairs (x, y) where $x \in M$ or $x \in N$ and $y \in P$, while the second consists of all ordered pairs (x, y) where $x \in M$ and $y \in P$ or $x \in N$ and $y \in P$. Therefore they are equal. (See Figure 68.)

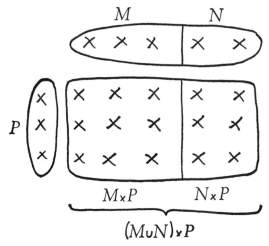

Figure 68

Since the two sets are equal, there is a bijection from one to the other: the identity map will do. So the sets are equinumerous, the numbers are equal, and the law is proved to be true.

Similarly one can prove the other laws of arithmetic.

To end this section, we'll put a new interpretation on the way a child counts a collection of objects. He points to each object in turn, and recites, 'One, two, three, ...' If as our standard sets for defining numbers we took $\emptyset$, $\{1\}$, $\{1, 2\}$, $\{1, 2, 3\}$, $\{1, 2, 3, 4\}$ (where 1, 2, 3, 4, are just symbols) then this counting looks suspiciously like setting up a bijection between the given set and one of our standard sets (Figure 69).

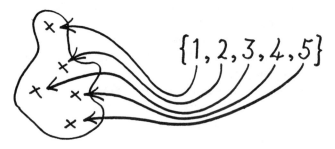

Figure 69

Infinite Arithmetic

Georg Cantor noticed that all of our previous discussion is valid for infinite sets as well as finite ones. (At no place did we specify that the sets involved were finite; but our examples were deliberately chosen to be finite.) The idea of a bijection makes good sense for infinite sets, so we can say what 'equinumerous' means; and we may then define infinite 'numbers' with the property that equinumerous sets have the same number, and conversely.

To avoid offending people's finer sensibilities we use a different word to describe these 'numbers': in fact we revamp an existing word. We call them *cardinals* (or transfinite numbers). For finite sets, the cardinal is the number of elements: for infinite sets it has many properties reminiscent thereof.

The definitions of addition and multiplication also make sense for cardinals; and the commutative, associative, and distributive laws hold. However, we have to pay some price for the extension into the infinite.

One curious property of cardinals was noted by Galileo, in 1638. There can exist a bijection between a set and a smaller subset of itself. The function f with $f(n) = n^2$ is a bijection between the set N of whole numbers and the subset of perfect squares. An infinite set can have the same cardinal as part of itself. Euclid's dictum 'the whole is greater than the part' must be amended to 'the whole is greater than or equal to the part' – where 'equal' really means 'equinumerous'.

A nice way to picture this bijection is to modify slightly the earlier diagrams for functions:

$$\begin{array}{ccccccccc}
0 & 1 & 2 & 3 & 4 & 5 & 6 & \ldots & n & \ldots \\
\updownarrow & \updownarrow & \updownarrow & \updownarrow & \updownarrow & \updownarrow & \updownarrow & & \updownarrow & \\
0 & 1 & 4 & 9 & 16 & 25 & 36 & \ldots & n^2 & \ldots
\end{array}$$

In a similar fashion there are bijections between N and the set of even numbers, or odd numbers, or integers, or primes:

$$\begin{array}{ccccccc}
0 & 1 & 2 & 3 & 4 & 5 & 6 & \ldots \\
\updownarrow & \updownarrow & \updownarrow & \updownarrow & \updownarrow & \updownarrow & \updownarrow & \\
0 & 2 & 4 & 6 & 8 & 10 & 12 & \ldots
\end{array}$$

$$
\begin{array}{cccccccc}
0 & 1 & 2 & 3 & 4 & 5 & 6 & \ldots \\
\updownarrow & \updownarrow & \updownarrow & \updownarrow & \updownarrow & \updownarrow & \updownarrow & \\
1 & 3 & 5 & 7 & 9 & 11 & 13 & \ldots
\end{array}
$$

$$
\begin{array}{cccccccc}
0 & 1 & 2 & 3 & 4 & 5 & 6 & \ldots \\
\updownarrow & \updownarrow & \updownarrow & \updownarrow & \updownarrow & \updownarrow & \updownarrow & \\
0 & 1 & -1 & 2 & -2 & 3 & -3 & \ldots
\end{array}
$$

$$
\begin{array}{cccccccc}
0 & 1 & 2 & 3 & 4 & 5 & 6 & \ldots \\
\updownarrow & \updownarrow & \updownarrow & \updownarrow & \updownarrow & \updownarrow & \updownarrow & \\
2 & 3 & 5 & 7 & 11 & 13 & 17 & \ldots
\end{array}
$$

So all these sets have the same cardinal. The cardinal of the set **N** of whole numbers is called $\aleph_0$ (aleph-zero). Cantor envisaged a whole system of infinite cardinals $\aleph_0, \aleph_1, \aleph_2, \ldots$, of which $\aleph_0$ is the smallest.

A set with cardinal $\aleph_0$ is said to be *countable* (because a bijection with **N** allows us to 'count' it – although we never stop counting!) Infinite sets which are not countable (and there are such sets) are called *uncountable*.

We have just shown that the sets A and B of even and odd numbers are both countable. They are disjoint, and their union $A \cup B$ is **N**. By the definition of addition for cardinals, the cardinal of **N** is the sum of the cardinals of A and B. But **N** is countable too, so we get

$$\aleph_0 + \aleph_0 = \aleph_0.$$

If we double $\aleph_0$ it stays the same size. This is another of the prices we pay for the inclusion of infinite sets. (Notice that we cannot deduce that $\aleph_0 = 0$, because we don't know how to subtract cardinals.)

Large and Small Infinities

We can compare cardinals with respect to size. For finite cardinals, if we have two sets M and N such that M has fewer elements than N, then we can find an injection from M to N (as in Figure 70).

We generalize this. If a and β are infinite cardinals, we say that a is *less than or equal* to β if there are sets A and B with cardinal a

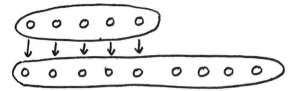

Figure 70

and β, such that there is an injection $f: A \rightarrow B$. In other words, some set with cardinal a can be paired off with a subset of a set with cardinal β. As usual we write

$$a \leq \beta.$$

Then we say that a is *less than* β if $a \leq \beta$ and $a \neq \beta$.

This order-relation on cardinals enjoys several pleasant properties:

    (i) $a \leq a$ for any cardinal a.

    (ii) If $a \leq \beta$ and $\beta \leq \gamma$ then $a \leq \gamma$.

    (iii) If $a \leq \beta$ and $\beta \leq a$ then $a = \beta$.

Property (iii) is far from easy to prove: it is known as the *Schröder–Bernstein theorem.* A proof can be found in Birkhoff and MacLane.[3]

So far the only infinite cardinal we know of is $\aleph_0$. Are there any others?

One might hope that the set of rational numbers, **Q**, had larger cardinal: after all, between any two integers there are infinitely many rationals. But this hope is not substantiated.

Imagine all the rational numbers p/q ($q \neq 0$) arranged in an infinite square array:

	$\vdots$	$\vdots$	$\vdots$	$\vdots$	$\vdots$	$\vdots$	
...	$3/-3$	$3/-2$	$3/-1$	$3/1$	$3/2$	$3/3$	...
...	$2/-3$	$2/-2$	$2/-1$	$2/1$	$2/2$	$2/3$	...
...	$1/-3$	$1/-2$	$1/-1$	$1/1$	$1/2$	$1/3$	...
...	$0/-3$	$0/-2$	$0/-1$	$0/1$	$0/2$	$0/3$	...
...	$-1/-3$	$-1/-2$	$-1/-1$	$-1/1$	$-1/2$	$-1/3$	...
...	$-2/-3$	$-2/-2$	$-2/-1$	$-2/1$	$-2/2$	$-2/3$	...
...	$-3/-3$	$-3/-2$	$-3/-1$	$-3/1$	$-3/2$	$-3/3$	...
	$\vdots$	$\vdots$	$\vdots$	$\vdots$	$\vdots$	$\vdots$	

Now imagine a spiral path through the array, starting at 0/1 (Figure 71).

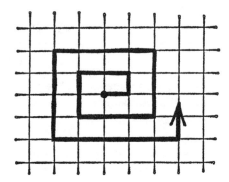

Figure 71

By going far enough along this path we reach any rational number p/q. We can define a function $f{:}\mathbf{N} \to \mathbf{Q}$ by: $f(n)$ is the nth *distinct* rational number along the path. The rule is unambiguous, so we have defined a function. It is surjective, because the path reaches *every* rational. By taking just distinct rationals we make f injective. So we have set up a bijection between $\mathbf{N}$ and $\mathbf{Q}$. Hence $\mathbf{Q}$ also has cardinal $\aleph_0$.

The first few rationals along the path are $0/1, 0/2, 1/2, 1/1, 1/-1,$ $0/-1, -1/-1, -1/1, -1/2, -1/3, 0/3, 1/3, 2/3, 2/2, \ldots,$. There are repetitions of value: $0/1 = 0/2 = 0/-3 = 0/3 = 0$, $1/1 = -1/-3 = 2/2 = 1$, etc. Eliminating these, the bijection goes

0	1	2	3	4	5	6	7	8 ...
$\updownarrow$	$\updownarrow$	$\updownarrow$	$\updownarrow$	$\updownarrow$	$\updownarrow$	$\updownarrow$	$\updownarrow$	$\updownarrow$
0	1/2	1	-1	$-1/2$	$-1/3$	1/3	2/3	2 ...

The pattern in the lower line is not obvious; but we know how to obtain it from the spiral. It would be hard to find a formula for the nth rational along the path, but since functions need not be defined by formulae this doesn't matter.

Another possible candidate for a cardinal bigger than $\aleph_0$ is the set $\mathbf{R}$ of real numbers. Since any real number can be approximated arbitrarily closely by rationals, one might expect the reals to have the same cardinal. But this time this is not the case. The cardinal of the set **R** – for the moment call it **c** – is larger than $\aleph_0$.

This is proved by contradiction. We assume that we can find a bijection between the whole numbers and the reals. Each real number will be of the form

$$A \cdot a_1 a_2 a_3 \ldots$$

where A is a whole number and each a_i is one of 0–9. With decimals we have to remember that there are ambiguities: $0 \cdot 100000 \ldots = 0 \cdot 0999999 \ldots$. These only occur for repeated 0s or 9s so we agree not to use repeated 9s. This makes our notation unambiguous.

A bijection **N** → **R** will look like:

$$
\begin{aligned}
0 &\leftrightarrow A \cdot a_1 a_2 a_3 a_4 a_5 \ldots \\
1 &\leftrightarrow B \cdot b_1 b_2 b_3 b_4 b_5 \ldots \\
2 &\leftrightarrow C \cdot c_1 c_2 c_3 c_4 c_5 \ldots \\
3 &\leftrightarrow D \cdot d_1 d_2 d_3 d_4 d_5 \ldots \\
4 &\leftrightarrow E \cdot e_1 e_2 e_3 e_4 e_5 \ldots \\
&\ \vdots \qquad\qquad \vdots
\end{aligned}
$$

where (in imagination) all the numbers in **N** appear down the left, and *all* the numbers in **R** down the right.

Now we exhibit a number in **R** that is *not* listed on the right. It will be of the form

$$0 \cdot z_1 z_2 z_3 z_4 z_5 \ldots$$

where we choose z_1 different from a_1, z_2 different from b_2, z_3 different from c_3, z_4 different from d_4, z_5 different from $e_5 \ldots$. In general z_n will be different from the nth decimal place of the number opposite $n-1$. To avoid ambiguities, we also choose the zs not equal to 0 or 9.

This is a real number. But it's not equal to the first in the list, because it differs in the first decimal place. It's not equal to the second, because it differs in the *second* decimal place. It's not equal to the third ..., and in general, it differs from the nth number in the list (which is in row $n-1$) in the nth decimal place.

So we have found a number not in the list.

But we started out, we claimed, with a complete list.

This is a contradiction. The only possibility is that no such list exists, so that there is *no* bijection $\mathbf{N} \to \mathbf{R}$. This tells us that the cardinal $\mathbf{c}$ of $\mathbf{R}$ satisfies

$$\mathbf{c} \neq \aleph_0.$$

But there is an obvious injection $\mathbf{N} \to \mathbf{R}$ (the identity on $\mathbf{N}$) so that

$$\aleph_0 \leq \mathbf{c}.$$

Combining these,

$$\aleph_0 < \mathbf{c}.$$

So in the sense of cardinals, there are more real numbers than rationals. But there are just as many rationals as integers.

So now we have a new cardinal $\mathbf{c}$ bigger than $\aleph_0$. We might wonder if $\mathbf{c}$ is Cantor's $\aleph_1$. This would be the case if there is no cardinal smaller than $\mathbf{c}$ but larger than $\aleph_0$. This problem was solved by Cohen in 1963, but the solution is so unexpected that we shall wait until Chapter 20 before saying what it is.

There are cardinals larger than $\mathbf{c}$. In fact, there is no largest cardinal: given any cardinal a we can find a larger one.

Take any set A with cardinal a. Let P be the set of *all* subsets of A. Let β be the cardinal of P. Then we can show that $\beta > a$.

First note that $f(x) = \{x\}$ defines an injection $A \to P$. So certainly $a \leq \beta$. If we had $a = \beta$ then there would be a bijection $h : A \to P$. For each $x \in A$ the element $h(x)$ is a subset of A; either $x \in h(x)$ or $x \notin h(x)$. We define a set T by

$$T = \{x \,|\, x \text{ does not belong to } h(x)\}.$$

Now T is a subset of A, so that $T \in P$. Since h is a bijection there is some $t \in A$ such that $h(t) = T$.

We ask whether or not $t \in h(t)$. If so, then $t \in T$. But for any $x \in T$, we know that $x \notin h(x)$, from which it follows that $t \notin h(t)$. On the other hand, if $t \notin h(t)$ then t passes the entrance requirement for T, so that $t \in T$. But $T = h(t)$ so $t \in h(t)$.

Either way gives a contradiction. So our assumption that h exists must be false. Therefore $\beta \neq a$, and all that remains is

$$a < \beta.$$

We could use this to give a different proof that $\mathbf{R}$ is uncountable. To each subset S of the integers we can associate a real number

$$0 \cdot a_1 a_2 a_3 \ldots$$

where $a_n = 1$ if $n \in S$, $a_n = 2$ if $n \notin S$. Different choices of S give different numbers. So we have defined an injection from the sub-sets of **N** into the set **R**. The set of subsets of **N** has cardinal larger than $\aleph_0$, and so **R** also has larger cardinal.

Transcendental Numbers

If all that could be done with infinite cardinals was proving theorems about infinite cardinals, nobody would have been very impressed by the idea. What forced mathematicians to take notice of them was the possibility of using them to prove theorems *not* about cardinals.

Some real numbers satisfy polynomial equations
$$a_n x^n + a_{n-1} x^{n-1} + \ldots + a_0 = 0$$
where the coefficients a_i are integers. For example, $\sqrt{2}$ satisfies
$$x^2 - 2 = 0.$$
Such numbers are said to be *algebraic*. Any real number that is not algebraic is said to be *transcendental*.

In Chapter 6 we remarked that all constructible numbers satisfy such an equation, with rational coefficients: by multiplying through by the product of the denominators of the coefficients we can get *integer* coefficients. So every constructible real number is algebraic. We also asserted that π satisfies no such equation: in other words π is transcendental.

For many years mathematicians suspected that π was transcendental, but they could not prove it. Worse, they could not prove that *any* numbers were transcendental. Then, in 1844, Liouville found a proof that such numbers did exist. But it wasn't a particularly easy proof. In 1873 Hermite proved that the number e (the 'base' of natural logarithms) was transcendental. Lindemann did the same for π in 1882.

But in 1874 Cantor found a very simple way to prove that transcendental numbers existed, *without actually finding any*. He used infinite cardinals.

Given a polynomial
$$a_n x^n + a_{n-1} x^{n-1} + \ldots + a_0$$
we define its *height* to be
$$|a_0| + |a_1| + \ldots + |a_n| + n.$$

For example, the height of the polynomial $x^2 - 2$ is
$$|-2| + |1| + 2$$
$$= 2 + 1 + 2$$
$$= 5.$$

Every polynomial, with integer coefficients, has finite height. More interestingly, there are only a finite number of polynomials with a given height h. Because the degree n must be $\leq h$, and there are only the possibilities $-h, -h+1, \ldots, -1, 0, 1, 2, \ldots, h$ for each coefficient; making at most
$$(2h+1)^{h+1}$$
polynomials of height h. (It would be possible to give better estimates than this, but for our purposes *any* estimate will do.)

We may therefore write out all possible polynomials with integer coefficients in a sequence: first we list those of height 1 (in any order), then those of height 2, then height 3, and so on. Because there are only finitely many polynomials of each height, the sequence does not get stuck forever at some fixed height, and so all polynomials appear somewhere in the sequence.

The polynomials of height 1 are just 1 and -1. Those of height 2 are 2, -2, x, $-x$. Those of height 3 are $2x$, $-2x$, $x+1$, $x-1$, $-x+1$, $-x-1$. So the sequence starts off:
$$1, -1, 2, -2, x, -x, 2x, -2x, x+1, x-1, -x+1, -x-1, \ldots.$$
Let the nth polynomial in this sequence be $p_n(x)$, so that the sequence now takes the form
$$p_1(x), p_2(x), p_3(x), \ldots, p_n(x), \ldots.$$
Every polynomial with integer coefficients occurs in this sequence.

The algebraic numbers are precisely the roots of the equations
$$p_n(x) = 0.$$
If the degree of the polynomial $p_n(x)$ is d, there are at most d roots of this equation. So we can arrange the roots in a sequence
$$a_1, \ldots, a_d.$$
If we fit all these short sequences together, one for each polynomial $p_n(x)$, we get a new sequence containing every algebraic number:

$$\underbrace{\beta_1, \ldots, \beta_i,}_{\substack{\text{roots of} \\ p_1(x) = 0}} \underbrace{\beta_{i+1}, \ldots, \beta_j,}_{\substack{\text{roots of} \\ p_2(x) = 0}} \ldots, \underbrace{\beta_k, \ldots, \beta_l,}_{\substack{\text{roots of} \\ p_n(x) = 0}} \ldots.$$

Of course, any given algebraic number may occur more than once in this sequence.

Now, for any whole number m we define $f(m)$ to be the $(m+1)$-th distinct algebraic number along the sequence. This makes f a function from N to the set of algebraic numbers. Because *every* algebraic number occurs in the sequence, f is surjective. By choosing *distinct* algebraic numbers, we make f injective. So f is a bijection, which means that the set of algebraic numbers has cardinal $\aleph_0$. The algebraic numbers form a countable set.

But we know that the real numbers form an *un*countable set. So some real numbers are not algebraic. This proves the existence of transcendental numbers.

In brief: transcendental numbers must exist, because there are more real numbers than algebraic ones.

This is a pure existence proof. It does not tell you one single transcendental number. It gives no clue, for example, as to the status of π. What it does is show that it is impossible for there not to be transcendental numbers.

In fact it shows that there are more transcendental numbers than there are algebraic. Because if there were only $\aleph_0$ transcendentals, the fact that every real number is either algebraic or transcendental would mean that

$$\aleph_0 + \aleph_0 = c$$

where c is the cardinal of R. But we already know that $\aleph_0 + \aleph_0 = \aleph_0$, which is *not* equal to c.

Prior to Cantor's theorem, mathematicians had become accustomed to thinking of transcendental numbers as being very rare, because they seldom seemed to use any. It came as a considerable shock to discover that they are extremely common: that *almost all* real numbers are transcendental. If you could pick a real number at random, you would be virtually certain to pick one that was transcendental.

Chapter 10   Topology

One of the most unexpected developments in twentieth-century mathematics has been the meteoric rise of the subject known as *topology*. Topology is sometimes described as 'rubber-sheet geometry', a whimsical and somewhat misleading description which nevertheless succeeds in capturing the flavour of the subject. *Topology is the study of those properties of geometrical objects which remain unchanged under continuous transformations of the object.* A *continuous* transformation is one in which points 'close together' to start with are 'close together' at the end of the cycle of transformation, such as bending or stretching. Tearing or breaking are not allowed. (There is, however, one *caveat*: since we are talking about transformations we are not interested in what happens anywhere except the beginning and the end. It is therefore permissible to introduce a break at some point provided it is eventually joined up again in the same way. For example we can untie a knot by cutting the string, undoing the knot, and then making the string whole again. This is why the 'rubber-sheet geometry' description is misleading.) It is possible to set up the definition of 'continuous' in a precise fashion, but we shall stick to the intuitive idea here. The question is raised again in Chapter 16.

What sort of properties are topological? Not the usual ones studied in Euclid's geometry. Straightness is not a topological property, for a line may be bent and stretched until it is wiggly. Neither is the property of being triangular: a triangle can be continuously deformed into a circle (Figure 72).

So in topology triangles and circles are the same thing. Lengths, sizes of angles, areas – all these can be changed by continuous transformations, and must be forgotten. Few, if any, of the customary concepts of geometry, remain in topology and new ones must be sought. This makes topology hard for the beginner until he gets the idea.

An archetypal topological property is that displayed by a certain kind of doughnut: having a *hole*. (Not the least subtle aspect of the matter is the fact that the hole is *not* a part of the doughnut.) No matter how one continuously distorts the doughnut, the hole

Figure 72

remains. Another topological property is the possession of an *edge*. The surface of a sphere has no edge, that of a hollow hemisphere does; and no continuous transformation will alter this.

Because continuous transformations are so varied, topologists make fewer distinctions. Anything having one hole will be much the same as anything else having one hole (as the next section illustrates). The topologist consequently has fewer objects to look at. The subject-matter is *simpler* than in most other branches of mathematics (although the subject itself is not). This is one of the reasons why topology has become a powerful tool, with applications throughout the whole of mathematics: its simplicity and generality make it widely applicable.

Topological Equivalence

The basic objects studied in topology are called *topological spaces.* Intuitively we should think of these as geometrical figures. Mathematically they are *sets* (often subsets of Euclidean space) endowed with some extra structure called a *topology* which allows us to set up an idea of continuity. The surface of a sphere, of a doughnut (more properly a *torus*) or of a double torus are all topological spaces (Figure 73).

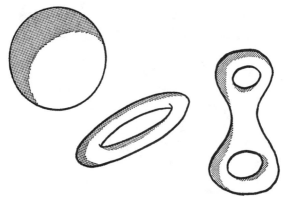

Figure 73

Two topological spaces are *topologically equivalent* if we can pass from one to the other in a continuous way, and also come back in a continuous way. The oft-quoted assertion that to a topologist a doughnut is the same as a coffee-cup provides an example (Figure 74).

Figure 74

In terms of set theory we start with two topological spaces A and B, and ask for a function $f:A \rightarrow B$ such that

(i) f is a bijection;
(ii) f is continuous;
(iii) the inverse function to f is also continuous.

The reason for wanting both f and its inverse to be continuous is the following. If we take two separate lumps and squash them together we get a continuous transformation (Figure 75) because

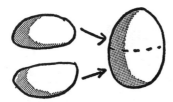

Figure 75

points originally close to each other remain so. But the inverse transformation takes one lump and pulls it into two separate pieces (Figure 76) so is not continuous, because points close to each other but on opposite sides of the dividing line end up further apart.

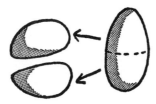

Figure 76

As an exercise try to classify the topological spaces shown in Figure 77 into topologically equivalent types.[1]

Some Unusual Spaces

If all topological spaces were as nice as the sphere and torus there would be little of interest in topology. A few instances of more exotic behaviour may help to tickle your intuition.

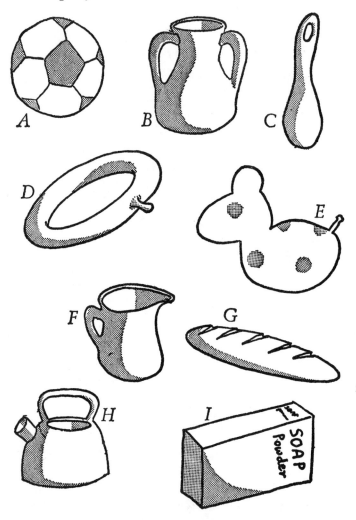

Figure 77

You have probably heard of the *Möbius strip* (or band), which can be made by taking a strip of paper and joining the ends with a 180° twist as in Figure 78.

This is topologically distinct from an untwisted cylindrical strip. It has exactly one edge. (Count it.) Since the number of

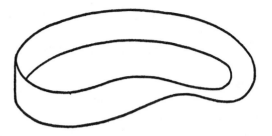

Figure 78

edges is a topological property, and since the cylindrical strip has two edges, the two strips are topologically inequivalent.

A more famous property of the Möbius strip is that it has only one side. A cylindrical strip could be painted red on one side, blue on the other. If you try to do this with a Möbius strip the two colours will run into each other somewhere.

Unfortunately it is difficult to make the one-sidedness mathematically respectable in a natural way. The strip has no thickness: every point of it is 'on' both sides, in the same way that every point of the plane is 'on' both sides. For the purposes of topology we must consider the strip as a space in its own right, rather than as a subset of Euclidean space, and it then becomes hard to see whether the number of sides is a topological property.

To make this clearer, let me ask a question: how many sides does three-dimensional Euclidean space have?

Most people, I think, would answer 'none'. It just goes on and on in every direction: how can it have any sides?

But suppose now that you are a flatlander, living in the two-dimensional plane, unaware of anything else. How many sides does your 'space' have? If you answered 'none' before then you must surely answer 'none' again: the plane just goes on and on in all directions.

In other words, the number of sides you are aware of depends on whether you think of the plane in its own right, or as a part of three-dimensional space. The same is true of three-dimensional space: if we think of time as a fourth dimension it has two sides – the past, and the future.

I hope you can see that it is now beoming rapidly more difficult even to define what we mean by the number of sides, let alone to see if this is a topological property.

However, there is a phenomenon which a hypothetical race of creatures living on a Möbius band could observe without bringing in any external considerations, and which provides a useful mathematical substitute for 'one-sidedness'. Assume that these creatures have two hands, with the thumbs pointing in different directions; they will have the concept of 'left' and 'right'. Furthermore let us suppose that they wear mittens (Figure 79).

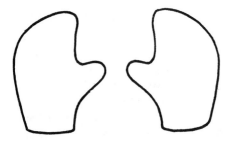

Figure 79

One cold morning one of our creatures awakes, to find that he has mislaid all of his right-hand mittens and has only left-hand ones. Being a resourceful beast, he takes one mitten and transports it around the strip, as in Figure 80.

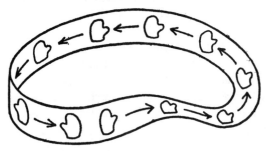

Figure 80

Much to our surprise, though not to his, it turns into a right-hand mitten. Of course, his right hand has become a left hand and his left hand a right: but in any case he now has a serviceable pair of mittens.

You can check this property by drawing on a paper Möbius band. But to take account of points being 'on' both sides of the paper it is necessary to look *through* the paper, either by holding it up to the light or by using a strip of transparent plastic in place of paper. Alternatively you can check it using your two hands and an imaginary Möbius band. Since your hands are not two dimensional, concentrate only on their *outlines*. Hold them in front of you, with the palms facing away, thumbs side by side, and fingers pointing upwards. Leaving the left hand fixed you must now move the right hand around an imaginary Möbius band, as follows. Lift your right elbow, tilting the right palm to the horizontal. Rotate the right thumb down and away from you until the hand is on edge with the thumb at the bottom. Lift the elbow still higher until the fingers point downwards and continue to rotate the hand in the same direction as before until the back of the hand faces you. Keeping the elbow raised and the fingers pointing down, with the back of the hand towards you, move the whole hand to the left until it is level with your left hand and on the far side of it. Now – and this requires a supple wrist – turn the thumb *away* from you until the right hand is edge on. Ideally now you should continue to rotate the right hand, but this is not anatomically feasible. Instead, rotate the *left* thumb towards you, until the two hands are side by side with the little fingers together: the left one pointing up and the right down.

You have now reached a very uncomfortable position which corresponds to having moved your right hand once around a Möbius band (and the left round the other way a little to meet it). For extra comfort, keep the two hands in the same position relative to each other, but move the right hand back a little towards the right and let the left hand follow it round towards the centre and away from you. Now you must turn the right hand upside down by rotating it on the surface of the Möbius band. To do this, bring the right elbow down to your side, keeping the right palm facing away from you. You should now have both hands pointing upwards, side by side, with the left palm facing towards you and the right palm away from you. Finally, move the two palms together: the hands fit neatly with the thumbs overlapping. As far as the *outlines* of the hands are concerned, you have now turned your right hand into a left hand by moving it around a Möbius band. (You also have a magnificent example of the style of mathematics known in the trade as 'handwaving'.)

A creature living on a two-sided world (this includes us, as far as we can tell at the present time) cannot perform the mitten trick.[2] To him, left and right cannot be interchanged. But to a Möbius beastie, the very concepts of left and right have a significance only if objects are not moved around. It is not possible to define leftness and rightness consistently over the whole Möbius band. It is said to be *non-orientable*. A space such as ours, in which a global definition of left and right can be made, is said to be *orientable*. Orientability corresponds to two-sidedness and non-orientability to one-sidedness; and both are intrinsic topological properties independent of any external space.

If two Möbius strips are joined edge to edge the result is known as a *Klein bottle* (Figure 81).

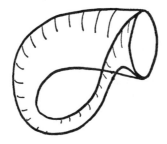

Figure 81

It has no edges, and is non-orientable because the Möbius bands are non-orientable. However, it cannot be embedded in space of three dimensions without crossing itself.

Another way to describe it is to imagine a square, whose edges are glued together as in Figure 82, so that corresponding arrows overlap. (First join the top and bottom, getting a cylinder. To fit the ends together the right way round, bend the cylinder and push it through itself.) You can use this diagram to check that it really is two Möbius bands edge to edge: cut it up again as in Figure 83.

Statements (frequently made) about the inside and outside of a Klein bottle are nonsense. It cannot be constructed in three-

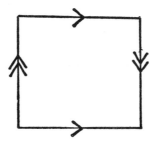

Figure 82

dimensional space. In four-dimensional space (which we discuss in Chapter 14) it can be made without any crossings; but now the ideas of 'inside' and 'outside' make no more sense than they

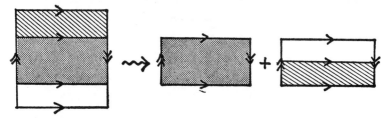

Figure 83

do for a circle in 3-space: one can pass from one to the other without encountering any obstacles.

Two other interesting spaces can be obtained by gluing up the edges of a square: the torus, and the *projective plane* (so named because of connections with projective geometry), shown in Figure 84.

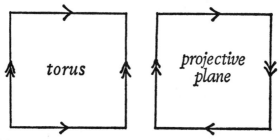

Figure 84

The torus we already know about: we note that it is orientable. The projective plane, like the Klein bottle, has no edges, is non-orientable, and won't embed in 3-space.

The projective plane is a Möbius strip and a disc sewn edge to edge. To do this in 3-space we have to twist the Möbius strip around to get its edge circular, and it then intersects itself, forming a *cross-cap* (Figure 85).

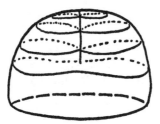

Figure 85

The projective plane is a cross-cap with the hole filled in (Figure 86).

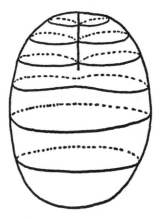

Figure 86

As a final curiosity, there is the *Alexander Horned Sphere* (Figure 87).

This is made by pushing out two horns from a sphere, twisting them together, splitting the ends, twisting these together, splitting

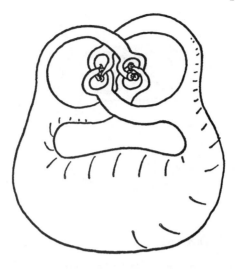

Figure 87

again, twisting again, and so on indefinitely; getting smaller and smaller at each stage. It is topologically equivalent to a sphere, believe it or not: the way the horns are pushed out can be used to define a suitable function. However, the space outside it is *not* topologically equivalent to the space outside an ordinary sphere.

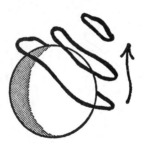

Figure 88

For outside an ordinary sphere any loop can be slid off (Figure 88), whereas for a horned sphere the loop may get entangled in the horns (Figure 89). Once again the trouble is caused by the surrounding space rather than the sphere itself.

Figure 89

The Hairy-Ball Theorem

Those are a few of the concepts and objects studied by topology: now we'll look at a theorem.

If you look at the way the hairs lie on a dog, you will find that they have a 'parting' down the dog's back, and another along the stomach. Now topologically a dog is a sphere (assuming it keeps its mouth shut and neglecting internal organs) because all we have to do is shrink its legs and fatten it up a bit (Figure 90).

Figure 90

One might wonder whether it is possible to comb the hairs in such a way that all partings were eliminated. This would give a smooth hairy ball, with none of the arrangements of hairs shown in Figure 91.

Figure 91

This is a question for topology, for if we deform the ball continuously, a smooth system of hairs will stay smooth, and a parting will stay a parting. Using topological methods it can be shown (not easily) that no perfectly smooth system of hairs exists. (The problem is more properly posed as one about 'vector fields' on a sphere, but the intuitive idea of a hairy ball gives the right feel.) The best that can be done is to comb the hairs so that everything is smooth except at *one* point, as in Figure 92.

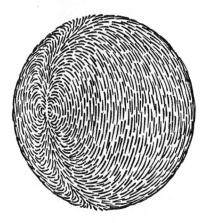

Figure 92

I don't propose to go into the proof. But the value of the result extends far beyond whimsical applications to hypothetical smooth dogs.

The surface of the earth is a sphere. The direction in which the surface winds are blowing, at any specific moment in time, gives a way of combing this sphere, with the flow-lines of the wind taking the place of hairs. The theorem says that no smooth system of winds exists (other than no wind at all which is impossible for other reasons), so that there must always be a cyclone somewhere.

Thus a knowledge of the *shape* of the earth allows us to draw conclusions about wind-patterns without any knowledge at all about the true behaviour of winds.

On a toroidal planet, stable smooth winds could exist, because a hairy torus *can* be combed smooth (Figure 93).

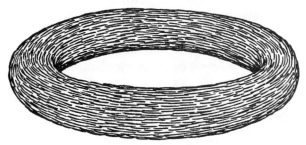

Figure 93

A deeper study, using more about winds, reveals that a more likely flow would be one winding around the torus as in Figure 94.

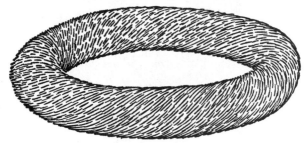

Figure 94

This is not the end of applications of the hairy-ball theorem. It has one application to algebra: it can be used to prove that every polynomial equation has solutions in complex numbers (the so-called 'fundamental theorem of algebra').

Chapter 11   The Power of
                Indirect Thinking

Ploughing straight ahead is not always the quickest way to make progress. It may be better to go round an obstacle rather than charge headlong into it. It is much the same in mathematics. Often a problem can seem insurmountable. You may even have a good idea of what the answer should be, but be unable to find any way of making certain. In these circumstances a fresh viewpoint, a new idea, can make all the difference.

How does one obtain a fresh viewpoint?

An explorer travelling through dense jungle usually has little idea of anything other than his immediate surroundings. If he comes to a mountain, he must climb it; a river, he must swim. But later, when people start to build roads through the jungle, they don't always follow the explorer's footsteps. They will have maps. They will say, 'Here is a mountain, there is a river,' and they will find ways to build the road round the mountain, and the best place to build a bridge over the river. If the explorer could have taken a more comprehensive view of the area – perhaps flown over it in a plane – he would have saved himself a lot of wasted effort: it might even have enabled him to succeed where he would otherwise have failed.

In mathematics it is too easy to concentrate very hard on one specific problem. If your methods are not good enough to handle the problem, you will eventually grind to a halt, baffled and defeated. Often the key to further progress is to stand back, forget about the special problem, and see if you can spot any general features of the surrounding area which might be of use.

Networks

There is an old puzzle which in one form concerns three houses, each of which must be connected up to supplies of water, gas, and electricity (Figure 95).

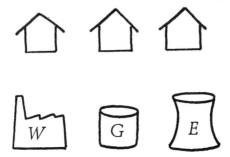

Can this be done in such a way that the connections do not cross each other, or pass through the houses or the sources of supply?

If you sit down with a pencil and paper, you will soon find that there is no apparent solution. But if you try to prove no solution exists, you run into a problem: there are so many possible ways of trying to draw the lines. Perhaps it would be helpful to have one of the lines loop around six or seven times first. It doesn't seem to be: but you can't see any way to show that it isn't.

This is exactly the sort of situation discussed in the introduction. The best way to proceed is to sit back and take a good look around.

The problem is not really about houses. It wouldn't matter if they were bungalows or blocks of flats. Nor would it matter if the source of electricity were next door, or a thousand miles away. Stripped of its picturesque language, the problem becomes that of starting with two sets, each of three points in the plane, and joining each point in the first set to each point in the second set with lines that do not cross.

Such questions come within the domain of mathematics known as *graph theory*, or the theory of *networks*.

A *network* has two main parts:

(i) a set *N*, whose elements are called *nodes* or *vertices*,

(ii) a way of specifying when two vertices are joined together.

We could make this abstract definition more respectable using set theory. But it is much easier to grasp the ideas involved by representing the vertices as dots, and joining them up by lines. The lines will be called the *edges* of the network. The precise disposition of dots and lines is not important; what *is* important is that the connections should be correctly made.

The diagrams in Figure 96 represent basically the same network. (Only the crossings marked with circles count.)

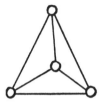

Figure 96

Each network has 4 vertices, with all possible pairs joined. One diagram can be turned into the other by moving dots and lines around, as in Figure 97.

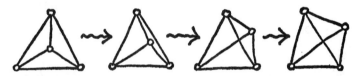

Figure 97

It is not even necessary to use straight lines; the same network is represented by the arrangements in Figure 98.

It is the *topological* structure of the network that is important.

Figure 98

It is not always possible to draw networks with no lines crossing. Networks which can be drawn without crossings are called *planar* networks – they can be drawn in the plane.

The puzzle with which we started can now be rephrased: *Is the network of Figure 99 planar?*'

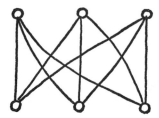

Figure 99

Before we can provide a satisfactory answer we must study the properties of planar networks.

Euler's Formula

A *path* in a network between vertices *a* and *b* is what one would expect: a sequence of edges, starting at *a* and ending at *b*, such that each edge ends where the next one begins. In Figure 100, *a* and *b* can be joined by a path.

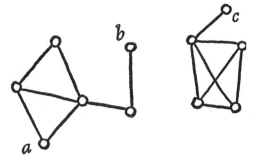

Figure 100

On the other hand, *a* and *c* cannot be joined by a path, because *c* lies in a part of the network which has no connections with the piece in which *a* lies. A network in which any two points can be joined by a path is said to be *connected*. That means that it does not fall into two (or more) distinct pieces. Any network is made up of connected pieces.

For this reason it is usual to confine attention to connected networks; more general ones can usually be dealt with by looking at the connected pieces.

The type of network we shall discuss from now on will be *finite* and *connected*. By 'finite' we mean that the number of vertices and edges is finite. Such a network is shown in Figure 101.

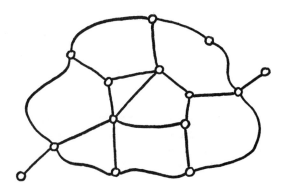

Figure 101

Any such network divides the plane into a number of regions, which we shall call the *faces* of the network. The one in Figure 101 has 8 faces. It has 14 vertices and 21 edges.

A network of the kind we are interested in (finite, connected, planar) resembles a map of an imaginary island; and to avoid cumbersome terminology we shall use the word *map* when referring to such a network.

At this point you should draw a few maps; and count the number of faces (*F*), vertices (*V*), and edges (*E*). Here are three to start you off (Figure 102).

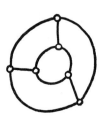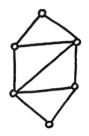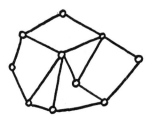

Figure 102

Tabulate your results like this:

F	V	E
8	14	21
4	6	9
4	6	9
6	10	15
...	...	...

Can you spot any connection between these numbers?

You might notice that E is always the largest. F and V are smaller, but between them come to much the same size: 22, 10, 10, 16 in the table above. These are all 1 larger than the corresponding E. So it looks as if

$$F+V = E+1$$

or

$$V-E+F = 1 \tag{†}$$

for any map.

The first person to prove that this formula does in fact hold for any map was Euler (1707–83). On the face of it, there is no very good *a priori* reason to expect *any* relation between F, V, and E; but anyone who took the trouble to count them for a dozen or so maps would eventually observe that (†) seems to be true. Not that this helps us to prove it.

With hindsight, there is a considerable clue in the expression $V-E+F$. It is the same for all maps. In particular, if we change one map into another one, it remains unaltered.

There are various ways in which $V-E+F$ can remain unchanged, but two of them are simpler than the others. The first is if E and F both decrease by 1. Then their difference doesn't alter, so that $V-E+F$ doesn't alter either. The second is if V and E both decrease by 1.

The first situation occurs if we erase from a map a face which is on the outside, together with one outside edge. If you like, we remove a stretch of coastline, together with one face, as in Figure 103.

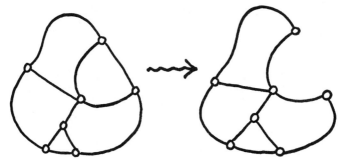

Figure 103

The second situation occurs if we have a single vertex 'dangling' at the end of an edge, as in Figure 104.

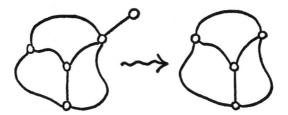

Figure 104

These operations are known as *collapses*. What we have noticed is that $V-E+F$ is unaltered by either type of collapse, and hence unaltered by any *sequence* of collapses.

Imagine the island to be surrounded by an angry sea. Bit by bit it erodes the coastline away, causing one or other kind of collapse. While this is happening $V-E+F$ remains serenely unchanged. The sea continues its work bit by bit . . . until there is no island left.

Did I say *no* island? That was hasty. Let's try it out (Figure 105)

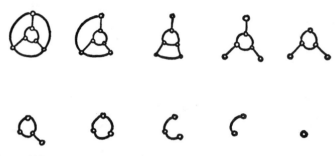

Figure 105

In fact, we reduce the island to a single point. 1 vertex, 0 edges, 0 faces. So for this, $V-E+F = 1-0+0 = 1$. But $V-E+F$ is not changed by collapses. So for the original island, we must also have had $V-E+F = 1$!

This, in flowery language, is essentially a proof. Every map can be collapsed to a point without changing $V-E+F$, and for a point this takes the value 1. So (†) holds for any map.

This formula – *Euler's formula* – is extremely useful in a surprising variety of situations. Our first application will be to the problem of the houses and the utilities.

Non-Planar Networks

The problem, once again, is: is the network of Figure 99 planar?

We apply what in chess language might be termed 'the mathematician's gambit': we concede that it might be planar. On the basis of this concession we try to deduce a contradiction: it will then follow that it *isn't* planar after all.

The network has $V = 6, E = 9$. We can't work out F because it isn't drawn on the plane. But if it were, then however it was drawn, it would have F faces, where

$$6-9+F = 1.$$

So we must have $F = 4$.

Now we have to work out where the faces will come, *without* actually drawing the network on the plane.

Each face of a planar map is surrounded by a closed *circuit*, or *loop*, of edges, as in Figure 106.

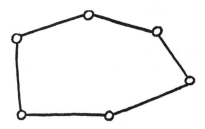

Figure 106

The perimeter ('coastline') forms such a loop. Now in Figure 99 the closed loops contain either 4 or 6 edges, as you should check. If it is drawn on the plane it will have 4 faces, each with either 4 or 6 edges; the different possibilities being

$$4 \quad 4 \quad 4 \quad 4$$

$$4 \quad 4 \quad 4 \quad 6$$

$$4 \quad 4 \quad 6 \quad 6$$

$$4 \quad 6 \quad 6 \quad 6$$

$$6 \quad 6 \quad 6 \quad 6.$$

Now let's count the edges a different way. Each edge lies on 2 faces, except for the ones round the outside. If we pretend that the outside is an extra very large face, it also has 4 or 6 edges; and we now have 5 faces˙

$$4 \quad 4 \quad 4 \quad 4 \quad 4$$

$$4 \quad 4 \quad 4 \quad 4 \quad 6$$

$$4 \quad 4 \quad 4 \quad 6 \quad 6$$

$$4 \quad 4 \quad 6 \quad 6 \quad 6$$

$$4 \quad 6 \quad 6 \quad 6 \quad 6$$

$$6 \quad 6 \quad 6 \quad 6 \quad 6$$

such that each edge touches 2 faces. So the sum of the edges of the faces is twice the number of edges altogether. In the above cases the number of edges must therefore be (respectively) 10, 11, 12, 13, 14, 15. But we already know that $E = 9$.

This is a contradiction, so the assumption of planarity was false. The network is *not* planar; and the 'mathematician's gambit' has payed off again. It follows that the puzzle cannot be solved using lines that do not cross, or go through houses or sources of supply.

The great thing about this proof is that it makes no mention of any possible ways of attempting to draw the connections. It by-passes all such considerations by admitting that there *might* be *some* way of making the joins, and then showing that there are none.

In a similar way we can deal with the network of Figure 107.

For this, $V = 5$ and $E = 10$, so F would be 6 if the network were planar. Now every face must have at least 3 edges. If we throw in

Figure 107

an extra face around the outside the sum of the number of edges of the faces is twice the number of edges: this sum is at least $3.7 = 21$, so the number of edges is greater than 10. But we know already that $E = 10$, a contradiction. So this network too is non-planar.

These two networks are important as prototypes for *all* non-planar networks. Kuratowski has proved that any non-planar network must contain within it one of these two. The result is not hard, but it takes several pages of case-by-case analysis, and we shall not go into the matter.[1]

The problem of planarity of a network has some practical applications to electronic circuits – particularly printed circuits and microminiature integrated circuits – but other considerations enter: the lengths of the connections may be important, and components can interfere with each other without actually touching.

Another Application

Euler's formula can be used to yield almost everything known about the famous (infamous?) *four-colour problem*: given a map, can it be coloured with 4 colours so that no 2 faces which touch along an edge have the same colour?

Certainly 4 are necessary (Figure 108), and it is possible

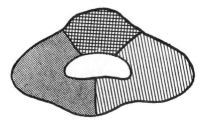

Figure 108

to show that no map can have 5 faces, each touching the other 4 (along an edge). However, this does not prove that 4 will always be enough.

The best that *is* known is that 5 colours will suffice. The gap between 4 and 5 has never been filled.* (Before you rush off to try, let me warn you that the problem is extremely subtle. I expect it to succumb only to a very deep understanding of planar networks.)

By 'map' I have so far meant a map on the plane, in our previous sense. The problem for a plane has the same answer as the analogous problem for maps on a sphere, because of what we might call the 'orange-peel trick'. Given a map on the sphere, we make a small hole inside one of the faces, whereupon it can be pulled open into a planar map. Conversely, a map on the plane can be folded around a sphere to give a map on the sphere, with the region outside the original map forming an extra face.

In consequence, if every map on the plane can be coloured with 4 colours, the same is true for the sphere, and vice versa. It happens to be more convenient to talk about the sphere, for the same reason that we introduced an extra face outside the map in the previous section. The extra face means that on a sphere we have

$$V - E + F = 2.$$

We shall now prove that 5 colours are enough to colour any map on the sphere (and hence on the plane). The proof will proceed as follows: we find ways to modify any given map which reduce the number of faces, so that from a 5-colouring of the reduced map we can reconstruct a 5-colouring of the original. By reducing often enough we get down to a map with 5 or fewer faces, which can obviously be 5-coloured; working backwards gives a 5-colouring of the original map.

(i) We can eliminate vertices where more than 3 faces touch. For if 4 or more do, some pair of them do not touch anywhere else, and we can merge these (Figure 109). If the

Figure 109

* It has now. See Appendix, p. 300.

new map can be 5-coloured, so can the old: we colour the 2 old regions the same as the single new one.

If the vertex we started with was surrounded by a large number of faces, we would have to reduce several times before getting down to only 3 faces.

(ii) We can remove 3-sided faces by merging them with a neighbour (Figure 110).

Figure 110

From a colouring of the new map, we can 5-colour the old by choosing a different colour for the 3-sided face from the 3 colours around it.

(iii) Similarly we can merge any 4-sided face with a neighbour (Figure 111). With 5 colours around, there is still one left to colour the merged face.

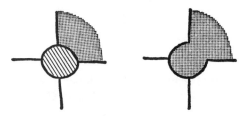

Figure 111

(iv) We now have a map on the sphere, all of whose faces have at least 5 edges. We prove that at least one face must have *exactly* 5 edges.

If the map has V vertices, E edges, and F faces, we know that each vertex lies on 3 edges (by step (i)). And each edge lies on 2 faces. So we have

$$3V = 2E = aF$$

where a is the *average* number of edges to a face. Since

$$V - E + F = 2$$

we have

$$\frac{a}{3}F - \frac{a}{2}F + F = 2$$

so that

$$a = 6 - \frac{12}{F}$$

which is *less than* 6. If the average number of sides per face is less than 6, then some face must have less than 6 edges. But every face has 5 or more edges. So there must be a face with exactly 5 edges.

(v) Consider such a 5-edged face P, having neighbours Q, R, S, T, U as in Figure 112.

Figure 112

Some pair of neighbours do not touch – and we can arrange our notation so that they are Q and S. Now merge all 3 regions P, Q, S (as in Figure 113).

If the resulting map can be 5-coloured, so can the original one: in the merged map Q and S have the same colour, so there are only 4 colours surrounding P, which leaves one spare.

(vi) Since the number of faces decreases with each merger, we eventually get a map with 5 or fewer faces. This obviously

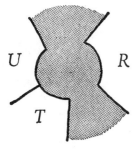

Figure 113

can be 5-coloured: all we do is pick a different colour for each face.

Reversing the steps, we find we can 5-colour the original map.

To recap: we apply a *reduction process* to the original map, gradually making it less complicated. We do this in such a way that at each stage we can colour the old map provided we can colour the new one. The proof ensures that eventually the new map *can* be coloured: so the previous one can, then the one before that ... and eventually we find out how to colour the original map.

To see how this works out in practice, draw yourself a map (not too many faces!), follow the reductions through, and find a way to colour it.

On surfaces other than the sphere the analogous problem has been completely solved. We discuss this briefly in Chapter 12. But for the sphere, the *simplest* kind of surface, all that is known at the present time* is that 5 colours suffice and 4 are necessary.

In a way it would be a pity if the problem *were* solved. It is a supreme example of a problem which is very easy to state, and incredibly hard to answer.

---

* No longer. See Appendix, p. 300.

Chapter 12   Topological Invariants

'The true traditional doughnut has the topology of a sphere. It is a matter of taste whether one regards this as having separate internal and external surfaces. The important point is that the inner space should be filled with good raspberry jam. This is also a matter of taste' – P. B. Fellgett

It is usually not too hard to prove that two given topological spaces are topologically equivalent (assuming this to be the case). All we need do is exhibit a suitable function between them.

Much harder is a proof that two inequivalent surfaces *are* inequivalent. We have to show that, out of a potentially infinite number of possible functions, no suitable one exists. The two spaces of Figure 114 (in which we are thinking of the surfaces, not the insides) are obviously topologically different. Yet how to prove it?

Figure 114

We can see that the torus has a hole in it, and the sphere does not. But the problem is that the hole is not in the torus at all, it's in the surrounding space. And we know that it's dangerous to draw conclusions that could depend on the surrounding space. As a topological space in its own right, the torus does not contain anything that we could call a hole. The hole is not a part of the torus that we see.

One way to distinguish between inequivalent topological spaces is to find some *topological* property which one has and the other hasn't. For example, every closed curve on a sphere divides it into two pieces (Figure 115), but on the torus there exist closed curves which do not cut it into separate pieces (Figure 116).

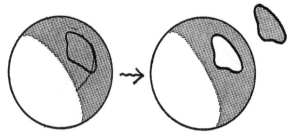

Figure 115

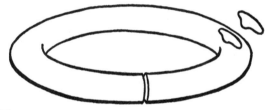

Figure 116

The properties: closed curve, connected, disconnected are topological; which proves that the two spaces, sphere and torus, are topologically distinct.

By refining this technique we could distinguish between, say, a surface with 19 holes and one with 18, but the details would be very messy and not terribly satisfying.

Euler's Formula Generalized

The property of the sphere, that for any map $V-E+F = 2$, is topological. Any continuous transformation applied to a sphere transforms a given map into one with the same values of V, E, and F.

If we try drawing maps on the torus, we find that $V-E+F$ is no longer 2. For the map in Figure 117 we have $V = 4$, $E = 8$, $F = 4$, so that

$$V-E+F = 0.$$

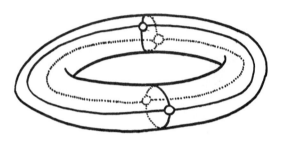

Figure 117

The same equation holds for any other map on the torus, for much the same reason that Euler's formula was true for any map on the plane or sphere. This property of maps on a torus is also topological.

We can generalize this formula to a wide range of topological spaces, known as surfaces.

Both the sphere and torus can be *triangulated*, that is, covered with triangles which fit together along their edges (Figure 118).

Figure 118

It doesn't matter that the triangles are not flat, or that the edges aren't straight: all we need are bits which are topologically equivalent to ordinary triangles.

Any space which can be built up from a finite number of triangles, so that any pair which touch do so along a single edge, or at a single vertex, is said to be *triangulable*. A *surface* is a topological space which is

  (i) triangulable,
  (ii) connected (i.e. all in one piece, as for networks),
  (iii) without any edges.

Examples of surfaces include the sphere, torus, Klein bottle, and projective plane: Figure 119 shows a triangulation of the projective plane.

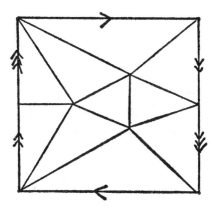

Figure 119

The Möbius strip is not a surface in this sense, because it has an edge. The plane is not a surface, because it cannot be built up from a *finite* number of triangles.

On any surface we can draw maps, as we did on the sphere, and we can count the faces, edges, and vertices. For a given surface S the number $V-E+F$ can be shown to be independent of which map we choose. It is known as the *Euler characteristic* of the surface, and is denoted by $\chi(S)$. Because it does not depend on the choice of map, it is the same for topologically equivalent spaces: in consequence it is a *topological invariant*. (A topological invariant is anything that is the same for topologically equivalent spaces.)

Another topological invariant is orientability. The torus cannot be topologically equivalent to the Klein bottle, because the torus is orientable and the Klein bottle isn't.

These two invariants: Euler characteristic and orientability, suffice to distinguish all the different surfaces we have so far encountered:

S	$\chi(S)$	orientable?
sphere	2	yes
torus	0	yes
double torus	−2	yes
projective plane	1	no
Klein bottle	0	no

Constructing Surfaces

Our eventual aim will be to classify all the possible surfaces, up to topological equivalence. The first step is to construct a set of standard surfaces.

The technique we use to construct them is known as *surgery*: cutting spaces up and then gluing them together again. It is very useful in topology.

Standard orientable surfaces are formed by sewing handles on spheres. If we sew on no handles, we get a sphere. To sew on a handle, we cut two holes out of the sphere and sew in a cylinder, whose edges are joined to the edges of the holes (Figure 120).

One handle gives a torus, two handles a double torus, and so on. The *standard orientable surface of genus n* is a sphere with n handles sewn on. (The word *genus* just gives us a peg on which to hang the number n.)

Standard non-orientable surfaces are obtained by sewing on Möbius strips. To do this, cut *one* hole in the sphere. This has a single circular edge: the Möbius strip also has a single circular

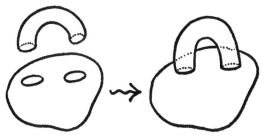

Figure 120

edge, and we join them up. If you try to do this in 3-space the Möbius band has to intersect itself, forming a cross-cap. But abstractly there is no cause for alarm (Figure 121).

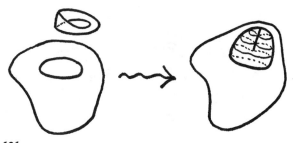

Figure 121

Sewing on one Möbius strip gives a projective plane (as in Figure 86), Sewing on two gives a Klein bottle, much as in Figure 83.

The Euler Characteristic of a Standard Surface

Our next step is to calculate the Euler characteristics of our standard surfaces. For the orientable case, we work as follows: we may assume that the two discs cut out of the sphere are faces of a map, part of which is shown in Figure 122.

For this map, $V - E + F = 2$, because we are on a sphere. Now, adding a handle as in Figure 123 changes this: we lose 2 faces on the sphere but gain 2 faces on the handle, the vertices do not

Figure 122

change, and we gain 2 edges. The net result is to *decrease* the Euler characteristic by 2. The same thing happens for each handle we add, so by adding *n* handles we decrease by $2n$. Therefore, the Euler characteristic of the standard orientable surface of genus *n* is

$$2-2n.$$

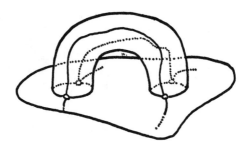

Figure 123

In particular, this proves that standard orientable surfaces of different genus are not topologically equivalent, because they have different Euler characteristics.

Next we turn to the non-orientable case. We can assume that the disc removed is part of a map looking like Figure 124.

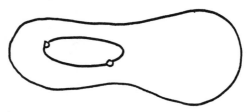

Figure 124

If we sew in a Möbius strip as in Figure 125 we find that we lose 1 face on the sphere, gain 1 face on the strip, and gain 1 edge on the strip. So this time the Euler characteristic decreases by 1 for each

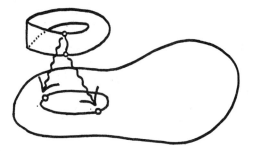

Figure 125

Möbius strip. For the standard non-orientable surface of genus *n* the Euler characteristic is

$$2-n.$$

Again, this suffices to distinguish between the standard non-orientable surfaces.

The two invariants, Euler characteristic and orientability, between them show that all our standard surfaces are topologically distinct. We shall now show that *any* surface is topologically equivalent to one of the standard ones.

Classifying Surfaces

The method of proof we shall use is due to E. C. Zeeman.[1] We use surgery to chop up a given surface into a number of pieces, which we can then reassemble into a standard surface. The chopping-up and reassembly will be done in a way which gives topological equivalence: we always join up pieces along the same lines that we cut them apart by.

Let *S* be a surface. We draw a curve on *S* which (if possible) does not split *S* into 2 pieces. If ever it transpires that we cannot find such a curve, we stop.

A narrow strip of surface on either side of the curve will be topologically equivalent to a strip with its ends joined. Therefore it is either a cylinder or a Möbius band.

Now we apply surgery. If the strip is a cylinder, we remove it and sew in two discs across the holes left. We mark each disc with an arrow to remind us how to put the cylinder back. If the strip is a Möbius band, we remove it and sew in *one* disc.

In the same way that we calculated the Euler characteristic of the standard surfaces (but in reverse) we see that every surgery increases the Euler characteristic; by 2 if we had a cylinder, by 1 if we had a Möibus strip. We now appeal to the

Unproved assertion A: the Euler characteristic of any surface is at most 2.

Thus our sequence of surgeries must stop, after finitely many stages. But it only stops if we can find no curve which does not split the surface up.

Unproved assertion B: a surface which is disconnected by every curve on it is topologically equivalent to a sphere.

So when our surgeries stop, we have a sphere.

Now we reverse the process. Three kinds of desurgery can happen.

(i) We have two discs with arrows going round in opposite directions, and sew in a cylinder. This is the same as sewing on a handle (Figure 126).

(ii) We have one disc, and sew in a Möbius band.

(iii) We have two discs with arrows going the same way:

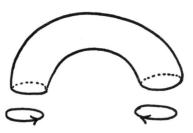

Figure 126

sewing a cylinder back is the same as sewing on a Klein bottle (Figure 127), which is equivalent to sewing on two Möbius strips (compare Figure 83). So we can convert the third kind of desurgery into two of the second kind.

Figure 127

If we started with an orientable surface S then only the first kind of desurgery occurs. So we end up with a sphere plus handles: which is a standard orientable surface. All we have done is pull S apart and put it back the same way (but keeping track of how the pieces fit) so S is topologically equivalent to the standard surface.

If we started with a non-orientable surface S, all three kinds can occur. We can eliminate the third kind as above. Because we are in the non-orientable case, at least one of the second kind must occur. If now we have any of the first kind, we can take one of the discs, and transport it around the Möbius strip. The result, as with the hypothetical mittens in Chapter 10, is to reverse the direction of the arrow on the disc. So now we have the third kind of desurgery; which we again convert into two applications of the second kind. So now we can desurger using only the second type of desurgery: sewing in a Möbius strip. But doing this gives us a standard non-orientable surface.

Apart from the two unproved assertions, we have shown that: *every surface is topologically equivalent to the standard orientable surface of genus $n \geq 0$ or to the standard non-orientable surface of genus $n \geq 1$.* (We don't need $n = 0$ in the second case, because that's just a sphere, which is orientable, and covered by the first case with $n = 0$.)

Assertions A and B were left unproved in order not to interrupt the flow. Now, we must deal with them.

The Proof of Assertion A

We can define the Euler characteristic of a network N to be
$$\chi(N) = V - E,$$
because a network as such has no faces. (Only when it is drawn on some surface will it be possible to define faces; in the above definition we therefore neglect them.)

If N is a network, we can show that $\chi(N) \leq 1$, as follows:

If N has any circuits, we break one, throwing away an edge only. This decreases E, and so *increases* $\chi(N)$. We can repeat this until no loops are left. A network without loops is called a *tree*, and looks like Figure 128.

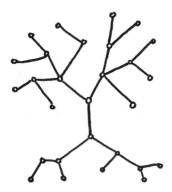

Figure 128

When we have a tree, we can use collapses (as in Chapter 11): we remove a vertex from the end of a 'branch', and also the edge attached. The result leaves the Euler characteristic unchanged. After enough collapses, we get down to a single point, for which the Euler characteristic is $1-0 = 1$. Going back to N: first we increased $\chi(N)$, then we left it unchanged, and eventually we got 1. So $\chi(N) \leq 1$.

We also see that the Euler characteristic of any tree is exactly 1.

Now we look at our surface S, and prove that $\chi(S) \leq 2$. We know that S is triangulable, so there exists a map on S (with triangular faces). We define a new map, the *dual map*, as in Figure 129: we put a vertex in the middle of each triangle, and draw an edge between any two vertices in adjacent triangles.

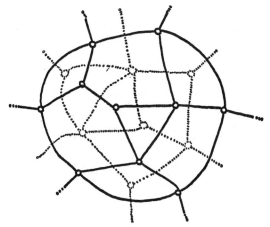

Figure 129

The vertices and edges of the dual map form a network. Inside this network we can find some trees (for example, a single point). From among these we take a maximal one: one that cannot be made larger without ceasing to be a tree. We call this a *maximal dual tree*. (An example is given by the heavy lines in Figure 130.)

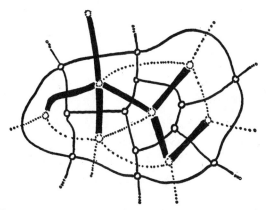

Figure 130

A maximal dual tree must contain all the vertices of the dual map. If not, we can connect up some new vertex by a path. This path will hit the dual tree at some point *P*, and the point *Q* before it hits will not be in the tree. Adding Q and the edge PQ to the tree still gives us a tree; but this contradicts maximality. So all the vertices must be included already.

Suppose that M is a maximal dual tree, and let C be that part of the original network comprising the original vertices, together with the edges that do *not* meet M. Then there are bijections between

(i) triangles on S and the vertices of M,

(ii) edges on S and edges of M and C (because each edge of S that is *not* in C crosses exactly one edge of M, by the way we defined C),

(iii) vertices on S and vertices of C (by the way we defined C: the vertices on S *are* the vertices of C).

This means that

$$\chi(S) = \chi(M) + \chi(C).$$

Now M is a tree, so $\chi(M) = 1$. And C is connected so $\chi(C) \leq 1$. Therefore

$$\chi(S) \leq 2$$

as claimed.

Proof of Assertion B

Let S be a surface, such that every closed curve on S disconnects S. We wish to show that S is a sphere.

First we prove that $\chi(S) = 2$. Let M and C be as before. Since

$$\chi(S) = \chi(M) + \chi(C)$$

it follows that, if $\chi(S) \neq 2$, then $\chi(C) \neq 1$. So C is not a tree.

Therefore C contains a closed loop. This is a closed curve on S, and by hypothesis disconnects S. But each piece into which S is split by the curve in C must contain a dual vertex. These have to be joined in M, so M must cut through the loop in C. But M and

C were defined to be disjoint. This is a contradiction; so our hypothesis that $\chi(S) \neq 2$ is wrong. So $\chi(S) = 2$.

Now it follows that $\chi(C) = 1$. So *C* is a tree. If you take a tree and 'fatten it up' a little, as in Figure 131, the result is topologically equivalent to a disc: all you need do is shrink the branches down towards some given point.

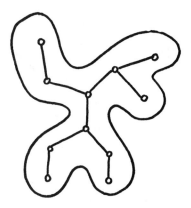

Figure 131

Define two subsets of *S* as follows: a point of *S* lies in *X* if it is nearer to *M* than to *C*. A point of *S* lies in *Y* if it is nearer to *C* than to *M*.

Each of *X* and *Y* is a 'fattening' of *M* or *C*, so is topologically a disc. Further, *X* and *Y* meet along their edges. So *S* is topologically equivalent to two discs sewn edge to edge . . . which is a sphere.

Map-Colouring on Surfaces

We can consider how many colours are needed to colour a map on a surface of standard type.

For Euler characteristic *n* it can be shown (by essentially the argument that showed 5 colours sufficient on a sphere) that
$$[\tfrac{1}{2}(7+\sqrt{(49-24n)})]$$
colours suffice, provided $n \leq 1$ (which is true except for the sphere).

For the torus, where $n = 0$, the formula gives 7. This is also a necessary number of colours, as shown by Figure 132.

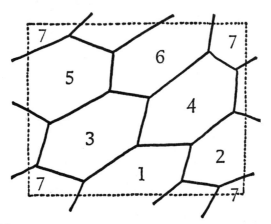

Figure 132

Recent research[2] has shown that the formula gives the exact number of colours required in all cases except two. For the sphere, it gives the answer 4, which may or may not be correct. For the Klein bottle it gives 7, which is definitely wrong: only 6 are needed.

This compounds the curious status of the 4-colour problem: *only* for the sphere, the simplest possible surface, do we *not* know the answer.*

* We do now. See Appendix, p. 300.

Chapter 13   Algebraic Topology

The Euler characteristic is a numerical invariant which, as we saw, can be used to distinguish between topologically inequivalent spaces. The search for other invariants has revealed a remarkable connection between two branches of modern mathematics: topology and abstract algebra. There are innumerable *algebraic* invariants associated with topological spaces: most commonly we associate a group with a space, in such a way that topologically equivalent spaces have *isomorphic* groups.

The hope is that by amassing enough invariants, we might be able to classify certain broad classes of topological spaces. Except for surfaces (where the Euler characteristic and orientability suffice) this has never been done, but mathematicians are closer than ever before to a genuine understanding of the problems involved.

Holes, Paths, and Loops

Suppose we wish to distinguish between a disc, and a disc with a hole in it.

We might notice that in the disc, any closed path can be shrunk until it reduces to a single point; but if there is a hole a closed path round the hole cannot be shrunk to a point – the hole gets in the way.

The 'shrinkability' of a closed curve is clearly a topological property, so we have achieved our immediate aim of distinguishing the two spaces. What I now want to do is develop the key idea: that holes can be detected by looking at paths in the space, and ways of deforming the paths.

Let's put our terms on a better footing. A *path* in a topological space is a line joining two points of the space. It doesn't matter how wiggly it is, or even whether it crosses itself; but there must be no breaks. We want the path to be continuous.

However, if the path does cross itself, we must specify which way to go around it: the two paths of Figure 133 are to be considered different.

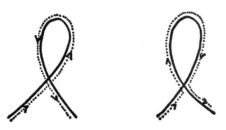

Figure 133

This is important: if we want to use paths to detect holes, the way we go round the path affects the way the path wraps around the hole. In Figure 134 one path can be 'pulled away from' the hole, and the other one can't.

Figure 134

The easiest way to specify how to travel along a path is to imagine a point moving along it. At time t this point will be in position $p(t)$. It starts at some time t_0 and ends at some time t_1. Since the path has no breaks, this means that p specifies a *continuous function* whose domain is the set of real numbers x in the interval $t_0 \leq x \leq t_1$, and whose target is the given topological space. Each such function defines a path, and each path defines such a function.

If one path ends where another begins, we can *compose* them by travelling along the first and then along the second as in Figure 135.

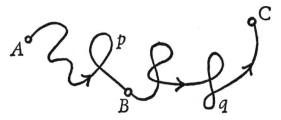

Figure 135

We use p to travel from A to B; then reset our clock to the starting time for q to travel from B to C. Let's denote the resulting path by

$$p^*q.$$

If p is defined on the interval $t_0 \leq x \leq t_1$ and q on the interval $t_2 \leq x \leq t_3$ then p^*q is defined on $t_0 \leq x \leq t_1 - t_2 + t_3$, because of the way we reset the clock in the middle.

Composition * of paths defines an operation on the set of paths: composing any two paths gives another, so the set is closed under the operation. Further, * is associative (Figure 136).

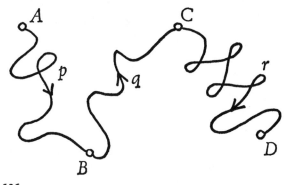

Figure 136

Going from A to B and then from B to C to D (i.e. $p^*(q^*r)$) is obviously the same as going from A to B to C and then C to D

(i.e. $(p*q)*r$)). (This is reminiscent of the reason why multiplication of functions was associative. But notice that $p*q$ is not the same as pq: in fact, the ranges and domains don't fit very well, and pq cannot be defined.)

However, we cannot compose *any* two paths: their ends have to fit. If we fix a *base point* A we can restrict attention to *loops* at A, that is, paths which start and finish at A. There is now no problem about composing loops, because the second always starts where the first one ends – at A. So the set of loops at A is closed under the operation *, and the associative law holds.

These are axioms (1), (2), and (3) for a group (see p. 100). Already we have produced some kind of algebraic structure. In fact axiom (4) holds as well; the trivial loop – stay at A and take no time in doing it – composes with any other loop to yield that loop.

The only thing that is missing is axiom (5), the existence of inverses. Now, what is an inverse? It is a way of *undoing* something. To undo a loop, we should travel along it in the opposite direction.

Unfortunately, this doesn't quite work. An inverse p^{-1} to a loop p should compose with p to give the trivial loop. But it takes no time at all to traverse the trivial loop: whereas to traverse $p*p^{-1}$ takes as least as long as traversing p.

We can't get round this by changing our choice of the identity: if $p*x$ is to equal p, then it must take no time to go round x, in order to leave long enough to go round p.

Nonetheless, we so nearly have a group that surely there is some way out?

Homotopy

The ring $\mathbf{Z}$ of integers does not possess multiplicative inverses. But if we chop it up into congruence classes to a prime modulus, inverses miraculously appear.

Our present predicament is analogous: we haven't got inverses but we want them. Its solution is similar: chop the set of loops up into some kind of classes, and operate on the bits.

What we need is something like 'congruence' for loops – although congruence in the sense of Euclidean geometry won't do. Our original purpose: using loops to find holes, gives a useful clue. In discussing the hole in the disc we talked about 'shrinking' the loop.

Given two loops in a space S we shall say that they are *homotopic* if one can be deformed continuously into the other inside S.

This time we really do want deformations and not just continuous functions. There is no problem about embeddings changing anything, because the loops are already embedded in S and S is what we are interested in.

In fact it is easier to illustrate homotopy for the more general paths. The definition is the same: but of course homotopic paths must have the same end-points. The two paths in Figure 137 are homotopic. (The deformation is shown by the sequence of dotted paths.)

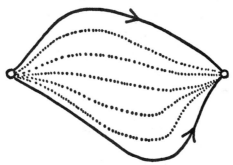

Figure 137

The paths of Figure 138 are not homotopic, because the hole gets in the way.

Instead of paths we consider *homotopy classes* of paths. Given a path p we let $[p]$ denote the set of all paths homotopic to p. This is the homotopy class of p, and it behaves in a way analogous with the congruence class of an integer.

If p^{-1} is the path p in reverse, then although $p*p^{-1}$ is not *equal* to the trivial loop, it is *homotopic*. As in Figure 139, we can shrink $p*p^{-1}$ gradually back towards the base point, at the same time going round it faster and faster. Eventually we get back to a path

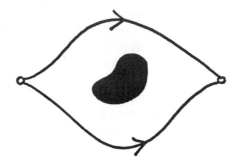

Figure 138

which stays at the base-point and takes no time in doing it. (For clarity we have slightly separated p and p^{-1}.)

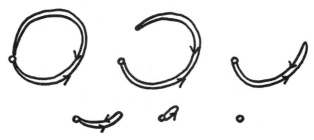

Figure 139

Now we've almost finished. We define composition of homotopy classes by

$$[p]^*[q] = [p^*q]$$

(and check that this makes sense). And we find that the set of homotopy classes of loops forms a *group* under the operation * of composition.

This is known as the *fundamental group* of the space S, and denoted by $\pi(S)$. Its construction is due to Poincaré.

If S and T are topologically equivalent spaces, we know there is a function $f:S \to T$ such that both f and its inverse g are continuous.

A continuous function turns paths in S into paths in T. The definition of composition of paths is topological, and so is the idea of homotopy: and f defines a function F such that

$$F([p]) = [f(p)]$$

on the homotopy classes. The way *F* is defined means that
$$F([p]*[q]) = F([p])*F([q]). \qquad (\dagger)$$
The inverse function $g: T \to S$ similarly defines a function G, which is the inverse function to *F*. So *F* is a bijection, and (†) says that *F* is an isomorphism.

Therefore the groups $\pi(S)$ and $\pi(T)$ are isomorphic. In this sense, $\pi(S)$ is a topological invariant.

You could extract from $\pi(S)$ *numerical* invariants, such as the order of $\pi(S)$, but in so doing you would lose useful information.[1]

The Fundamental Group of the Circle

The fundamental group is not much use unless we can calculate it. In general this is not an easy task, and together with its generalizations is the subject of a considerable body of theory.

For some spaces: **R**, $\mathbf{R}^2$, a disc, a solid ball, . . . , it *is* easy. These spaces have no holes, and any loop can be shrunk to give the trivial loop, as in Figure 140.

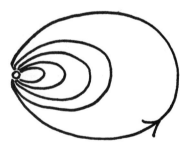

Figure 140

Their fundamental group is therefore the trivial group with one element *I*, such that $I^2 = I$.

Using the 'orange-peel trick' of p. 170 we can work out $\pi(S)$ when *S* is the surface of a sphere. Take any loop *p* on *S*. Choose a point not on *p*, and remove a small disc around this point, not meeting *p*. The remainder of the sphere can be opened up to a disc; inside this disc we can shrink *p* to a point. Folding the disc up again shows us how to shrink *p* to a point on *S*. So $\pi(S)$ is trivial, too.

The next simplest case is when *S* is a circle. Any loop in *S* winds around *S* a certain number of times. This number is called the *winding number* of the loop. The loops in Figure 141 (drawn

Figure 141

slightly separated for clarity) have winding numbers 1, 2, and 0 respectively. Reversing them gives winding numbers −1, −2, 0 (with the convention that anticlockwise ís the positive direction).

What we shall show is that the winding number determines the homotopy class: two paths are homotopic if and only if they have the same winding number.

This is intuitively reasonable. It seems hard to change the winding number just by deforming the loop. The trivial loop, of course, has winding number 0; and the third loop above, which also has winding number 0, can be shrunk back to a point.

To prove this, we introduce an extra space into the picture. It will have the advantage that its homotopy properties are easy to work out, and be sufficiently closely connected with the circle that we can deduce the homotopy properties of the circle.

Imagine a line *L* arranged over the circle, like a spiral staircase, with the point *O* of the line above our base point *A* in the circle. Any loop in the circle *S* can be 'lifted' to a path in the line *L*: imagine a point on *L* and a point on *S*. As the point on *S* moves round a loop, the point on *L* stays immediately above it, moving in a continuous fashion. For the loops of Figure 141 we get the paths of Figure 142.

The lifted path need not end up at *O*, although it always must end up directly above (or below) *O* on the spiral. The number of

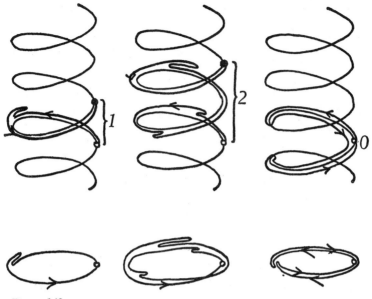

Figure 142

levels up or down it ends up at is exactly equal to the winding number. If you wanted, you could use this as a definition.

The crucial point now is that two paths in S are homotopic if and only if the lifted paths are homotopic in L. If we have a homotopy in L we can 'project' it down to give a homotopy in S. Conversely, any homotopy in S can be lifted to a homotopy in L: as we deform a path in S, we deform the corresponding path in L.

But in L homotopy properties are trivial, for L is a line, and we know that $\pi(L)$ is trivial. Two paths in L are homotopic if and only if they have the same end-points. This is certainly necessary; and it is sufficient because of the triviality of $\pi(L)$.

All of the lifted paths in L start at O. They have the same end-points if they end up the same number of levels above (or below) O. This happens precisely when the corresponding loops in S have the same winding number.

If we take a loop in S with winding number n and compose it with a loop with winding number m, we get a loop which goes first n times around, and then m times further. So it has winding number $n+m$. It follows that $\pi(S)$ is isomorphic to the group **Z** of integers under addition.

The Projective Plane

If we take S to be the projective plane, it turns out that $\pi(S)$ is the group with 2 elements

$$
\begin{array}{c|cc}
 & I & r \\
\hline
I & I & r \\
r & r & I
\end{array}
$$

The element r is the homotopy class of the path shown in Figure 143.

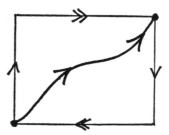

Figure 143

(Recall the way we think of the projective plane as a square with diametrically opposite points identified.)

The fact that $r^2 = I$ means that, although the path in Figure 143 cannot be shrunk to a point, the path obtained by going round it twice *can* be shrunk.

We can see this geometrically (Figure 144): we take one copy of the path and pull it across and over the top left-hand corner; because of the identifications it comes back in the lower right-hand corner in the reverse direction. Then we shrink the whole lot back to the base point.

This curious fact is connected with the 'soup-plate trick'. Obtain a soup-plate, preferably one that is not a family heirloom,

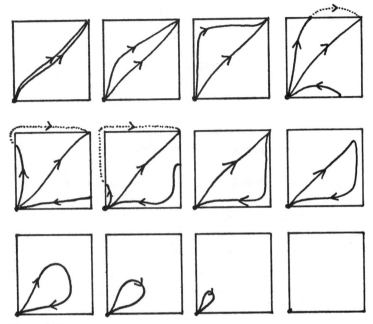

Figure 144

and hold it in front of you balanced on the tips of the fingers of your right hand. Bring your elbow back and down, passing the plate under your armpit. Keep twisting your arm the same way, and raise your elbow until the plate returns to its original position. Your arm is now twisted, and the elbow points upwards instead of down.

But you don't stop here. Continue to twist your arm the same way, moving the plate over your head and bringing your elbow around the front: and you will return to your starting position.

Halfway round, your arm was twisted. Going round again ought to twist it even more. But it doesn't: you end up in the original position with an untwisted arm.

This is exactly what is happening in the projective plane: going round once things get twisted; going round twice brings them back to normal.

Chapter 14   Into Hyperspace

*'People ask me, "Can you show us
this fourth dimension?." And I reply,
"Can you show me the first, second,
and third?" '* – Anonymous

Often a mathematical generalization, at first pursued for its own
sake, later turns out to be of great importance in mathematics as a
whole.

In Chapter 4 we found that the Euclidean plane can be thought
of as the set of all ordered pairs of real numbers, which we denoted by $\mathbf{R}^2$. In the same way, three-dimensional space can be
thought of as the set $\mathbf{R}^3$ of all triples (x, y, z) of real numbers.
And of course, the line $\mathbf{R}$ is one-dimensional. We have

1-space $= \mathbf{R} =$ the set of real numbers x
2-space $= \mathbf{R}^2 =$ the set of pairs of real numbers (x, y)
3-space $= \mathbf{R}^3 =$ the set of triples of real numbers (x, y, z)

Real space stops at this point. But reality is very disappointing.
Why not go on, putting

4-space $= \mathbf{R}^4 =$ the set of quadruples (x, y, z, u)
5-space $= \mathbf{R}^5 =$ the set of quintuples (x, y, z, u, v)

and in general

n-space $= \mathbf{R}^n =$ the set of n-tuples $(x_1, \ldots, x_n)$?

Why not, indeed? We are at liberty to make whatever definitions
we choose. But space does not consist of points alone. It has a
distance structure.

From Pythagoras it follows that the distance d between points
(x_1, x_2) and (y_1, y_2) is given by
$$d^2 = (x_1 - y_1)^2 + (x_2 - y_2)^2,$$
and the corresponding formula in 3-space is
$$d^2 = (x_1 - y_1)^2 + (x_2 - y_2)^2 + (x_3 - y_3)^2.$$
In 1-space we can even write the formula in a similar form,
$$d^2 = (x_1 - y_1)^2.$$

If everything is working properly, we ought to be able to define distance in 4-space by
$$d^2 = (x_1-y_1)^2+(x_2-y_2)^2+(x_3-y_3)^2+(x_4-y_4),$$
d now being the distance between (x_1, x_2, x_3, x_4) and (y_1, y_2, y_3, y_4). And in n-space one would expect the obvious formula.

It is not a question of whether the formula is *true*. We don't know anything about four-dimensional space, so we have no possible way to check the truth of the formula. We are setting up a piece of abstract mathematics; we may use whatever formula we wish. A more helpful attitude would be 'That's all very well, and I agree that it's the obvious formula to use, but can you do anything with it that makes sense?'

For something to qualify as a 'distance' it should satisfy three conditions:

  (i) The distance between any two points is positive.
  (ii) The distance between two points is the same in either direction.
  (iii) The distance from A to B should not be longer than the distance from A to C plus the distance from C to B.

Condition (iii) says that any side of a triangle is shorter than the other two put together, and corresponds roughly to the idea that 'a straight line gives the shortest distance between two points'.

We can verify these conditions for our formula.

Condition (i) holds, provided we take the positive square root to get d. And we can take the square root, because the right-hand side is a sum of squares, so is positive.

Condition (ii) says that if we change all the xs to ys and all the ys to xs, d should not change. But since $(x_1-y_1)^2 = (y_1-x_1)^2$, and so on, this is the case.

Condition (iii) leads to an important algebraic inequality. We'll take the special case $n = 2$. From Figure 145, it boils down to proving that
$$\sqrt{(a^2+b^2)}+\sqrt{(c^2+d^2)} \geq \sqrt{((a+c)^2+(b+d)^2)}.$$
Squaring, this will be true provided
$$a^2+b^2+c^2+d^2+2\sqrt{(a^2+b^2)(c^2+d^2)} \geq (a+c)^2+(b+d)^2,$$
i.e. provided
$$2\sqrt{(a^2+b^2)(c^2+d^2)} \geq 2(ac+bd).$$

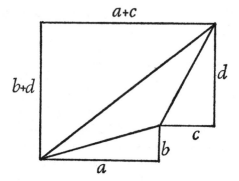

Figure 145

This, in turn, is true provided
$$(a^2+b^2)(c^2+d^2) \geq (ac+bd)^2$$
which is the same as
$$(a^2+b^2)(c^2+d^2)-(ac+bd)^2 \geq 0.$$
But you can work out the left-hand side, and it turns out to be equal to.
$$(ad-bc)^2.$$
But squares are always positive. So condition (iii) is true, at least in 2 dimensions.

Similar but messier calculations work for n dimensions. And already our tentative idea has generated an important inequality.

This means that, at the very least, our definition of distance is a sensible one. The geometers of the nineteenth century (who were rather running out of theorems to prove in 3 or fewer dimensions) began to investigate the properties of our abstract 4-space. To their great delight, they found that the concept was not only sensible, but very fruitful; full of beautiful ideas and theorems.

Polytopes

In 3-space there are exactly 5 regular polyhedra: tetrahedron, cube, octahedron, dodecahedron, icosahedron. The corresponding object in 4-space is called a *polytope*. Its 'faces' are 3-dimensional regular polyhedra, in the same way that the faces of a regular polyhedron are regular polygons; and the arrangement of 'faces' must be the same at each vertex. To avoid confusion, we'll call the 3-dimensional 'faces' *solids* and leave the word 'face' for the 2-dimensional faces of these solids.

The geometers (in particular Schläfli) found that there are exactly *six* regular polytopes in 4 dimensions:

solids	faces	edges	vertices	type of solids	name
5	10	10	5	tetrahedron	simplex
8	24	32	16	cube	hypercube
16	32	24	8	tetrahedron	16-cell
24	96	96	24	octahedron	24-cell
120	720	1200	600	dodecahedron	120-cell
600	1200	720	120	tetrahedron	600-cell

(We will return, later, to the pattern of the numbers.)

However, in 5-space, 6-space, or higher, there are only *three* regular polytopes: analogous to the tetrahedron, cube, and octahedron.

So the number of regular figures in 2-, 3-, 4-, 5-, . . . -space is $\infty, 5, 6, 3, 3, 3, \ldots$.

We can't draw any of the polytopes on paper. But no more can we draw 3-dimensional figures on paper. We use a convention to represent 3 dimensions on the two of the paper, the convention we use being determined by the anatomical structure of our eyes. We can represent 4-dimensional figures, too. But without practice the pictures will be hard to 'read' – in the same way that engineering drawings are hard for the non-engineer to cope with.

Four-Dimensional Pictures

One way to represent 4-dimensional figures is by *projection*. This is the way an artist draws a 3-dimensional scene on a 2-dimensional canvas. The scene is 'squashed flat' either radially or perpendicularly, as in Figure 146.

An analogous procedure enables us to project 4-dimensional figures into 3-space. An additional complication arises in printing:

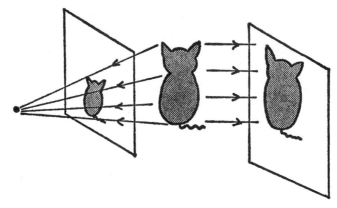

Figure 146

the 3-dimensional projection must itself be projected into 2-space!
Two projections of the hypercube are shown in Figure 147.

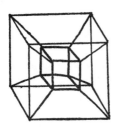 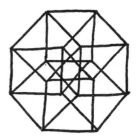

Figure 147

In interpreting these pictures you have to take account of the
effects of perspective. The small cube on the inside of the left-hand
figure is, in reality, the same size as thé outside one. But without
much trouble you can see that the hypercube *is* made up of eight
cubes (in the left-hand picture, one large, one small, and six dis-
torted into the shape of a truncated pyramid). Each cube is face
to face with six others, and there are four cubes around each
vertex.

There exist computer programs which display on a screen projections of 4-dimensional figures. The operator can control the
direction in which the projection occurs by 'rotating' the figure.

It is said that after some experience the operator can begin to guess how the projections will look as he rotates the figure – he begins to *think* in 4 dimensions. Topologists who work with higher dimensional spaces also tend to acquire this faculty.

There is another way to represent 4-dimensional objects, which seems to be easier to visualize: we draw a series of cross-sections. This is analogous to the way a map shows hills and valleys by drawing contours: an imaginary horizontal plane is allowed to slice the surface of the countryside, and the curves in which it cuts the surface are drawn for different positions of the plane, as in Figure 148.

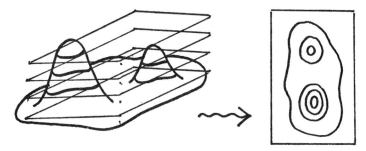

Figure 148

By cutting out pieces of card in the shape of the contours and stacking them on top of each other at the correct heights, you can reconstruct the shape of the surface.

A race of creatures living in 2-space could use these cross-sections to visualize 3-dimensional objects. And an inhabitant of 1-space could obtain an idea of the shape of a plane figure by taking a series of linear cross-sections.

In each case the sections drop a dimension. So the cross-sections of a 4-dimensional figure will be 3-dimensional.

By generalizing the algebraic formulation of taking cross-sections in 3-space, we can define exactly what we mean by a cross-section of an object in $\mathbf{R}^4$, or $\mathbf{R}^5$, ... We can use analogies to guess what these sections should look like in certain cases, and then check that we are right by algebra. *We* won't bother with the algebra.

Sections of a sphere form circles, growing from a point to a maximum and then contracting again. So sections of a *hypersphere* (the 4-dimensional analogue) should be spheres, growing from a point, and then contracting, as in Figure 149.

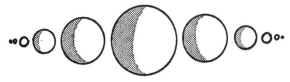

Figure 149

Sections of a cube are always square, so sections of a hypercube should be cubes (Figure 150).

Figure 150

Stacking the Sections

Our main problem, living in 3-space, is to stack the sections mentally. Again an analogy with hypothetical inhabitants of the plane helps. How could flatlanders stack 2-dimensional sections?

They could get them in the correct sequence by imagining that the plane doing the slicing moves uniformly in time. At a given time *t* the cross-section is a 2-dimensional object. If they could make a motion picture, whose successive frames corresponded to successive instants of time, they could use this to stack the pictures. They would think of a sphere as a circle, growing from a point, then shrinking.

In the same way, we can stack the 3-dimensional sections of a 4-dimensional object in time, thereby producing a 3-dimensional moving picture. A hypersphere would look like a bubble, growing and then shrinking. A hypercube would be a cube which suddenly appeared, stayed exactly the same for a period of time, and then suddenly disappeared. If you saw a sphere suddenly appearing, staying the same, and then suddenly disappearing, you would know that you were watching a hypercylinder with spherical cross-section.

This is better, but not good enough. As it stands, we have to watch the movie in a fixed sequence. We are like a blind man who is allowed to feel objects from top to bottom in a single movement. We want to be able to run our hands up and down, exploring closely any particularly puzzling (or interesting!) feature.

In short, we require a time-machine – or at the very least a reversible film-projector with variable speed.

The machine will have a foot-pedal to control time, and some sort of screen for displaying 3-dimensional pictures. Fortunately an imaginary machine will do: use your foot as a pedal and *think* of the pictures.

By varying the pressure of your foot you will be able to move around in time. As a first exercise in pedal-control, we shall undo a knot (in 4-space) without untying the ends. For simplicity a simple overhand knot will be used, but any other knot would do. Frame A in Figure 151 shows such a knot; it lives in 3-space at time $t = 0$.

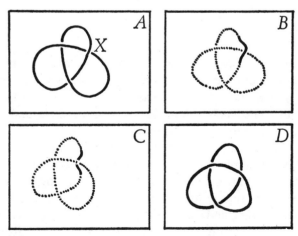

Figure 151

Grasp the string firmly at a point near the crossing X. Depress your foot, and move slightly forward in time, thereby dragging a small loop of string with you in the time direction, though leaving most of the knot in its original spatio-temporal state (shown dotted) as in frame B. Now push the loop downwards, below where the other bit of the string used to be, as in frame C. Finally, return to time $t = 0$, bringing the loop with you. The result (frame D) is to untie the knot.

As a second exercise, you might try to embed a Klein bottle in 4-space without self-intersections. Start with Figure 81 at time $t = 0$. Take hold of one bit of the 'tube' near the intersection, and move it slightly in time.

Another thing you can visualize is the linking of a circle and a sphere. First, let's set up an analogy. Consider two linked circles in 3-space; for convenience make one of them rectangular (Figure 152). Put in a time axis as shown.

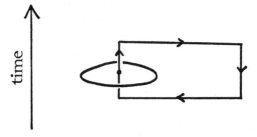

Figure 152

The link now looks like this: at time $t = 0$ a point starts in the centre of the rounded circle. It moves forwards in time, sideways in space, backwards in time to *before* the circle, sideways in space to a point immediately before the centre of the circle, and finally forwards in time to join up the loop.

For a link of a sphere and circle in 4-space, we do much the same. Imagine a sphere at time $t = 0$. A point starts in the centre of the sphere. It moves forwards in time (not cutting the sphere, because that disappears as soon as it starts to move), loops around the sphere in time and space to a point immediately backwards in time from the centre of the sphere, and finally moves forwards in time to close the circle.

Instead of links, you can try to visualize knots. It is possible to knot a sphere in 4-space, in the same way that a circle can be knotted in 3-space. Topologists have expended much effort on the problem of whether an *m*-sphere can be knotted in *n*-space. At the time of writing the first unsolved case is the 10-sphere in 17-space.

Astronauts in 24-Space

Imagine a pendulum swinging through a small angle. At any given time t it has position p and velocity q. If we plot a graph of p against q (with time measured in suitable units) we get a circle, and the point (p, q) moves around the circle with uniform speed as the pendulum swings (Figure 153). Thus, starting at A we have

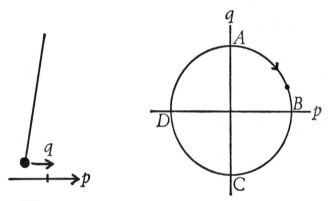

Figure 153

$p = 0, q > 0$; at B we have $q = 0$ and $p > 0$, at C we have $p = 0$ but now $q < 0$, and at D we have $q = 0$ and $p < 0$, which agrees with how a pendulum swings (Figure 154).

The diagram of p against q is known as a *phase diagram* and the (p, q)-plane is called *phase space*. In this case it has 2 dimensions because the state of the pendulum is determined by two numbers: one position coordinate and its velocity.

Any dynamical system has a corresponding phase space, with one dimension for each position variable and one for each velocity variable.

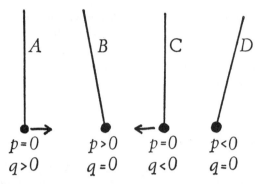

Figure 154

The system of sun, moon, and earth, acting by gravitational attraction, forms a dynamical system. There are three position variables for each body, and three velocity variables (because in 3-space we need three coordinates to fix positions or velocities), making a total of 18 dimensions for the phase space. The state of the whole system at any time is represented by a single point in phase space; as time varies this point describes a path, which completely specifies the motion of the whole system.

To compute the orbit of a spacecraft moving in the system, we have to throw in 6 more dimensions of phase space (for the craft), and the problem becomes one of 24-dimensional geometry! Not only is this a way of describing the problem. If developed systematically it gives a deep and powerful mathematical method: geometrical dynamics.

For a given dynamical system, there will be many ways to set the motion going. For the spaceship, we could choose various initial positions or velocities. To each initial state will correspond a point in phase space. As the system develops this point describes a path, so we get a family of paths, one for each initial position. If you imagine phase space to be filled with a fluid, such that each particle of fluid corresponds to a state of the system, then the fluid will flow along the paths. For the pendulum, the flow-lines will be concentric circles: the stationary point in the centre represents a pendulum hanging vertically at rest (Figure 155).

Coincidentally it follows from Newton's law of conservation of energy, that this imaginary fluid behaves exactly like a genuine

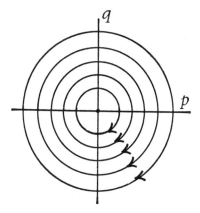

Figure 155

fluid; and further is incompressible. As a result, the methods of fluid dynamics can be applied to the general theory of dynamical systems. Without using multidimensional geometry, this application could never have been made.

Euler's Formula Generalized Further

Euler's formula gave a relation between the numbers of faces, edges, and vertices of a map in the plane. We have generalized this to other surfaces; now we ask whether there exists a generalization to spaces of higher dimension.

A 'map' in n-dimensional space will have n-dimensional regions, with $(n-1)$-dimensional faces; these in turn have $(n-2)$-dimensional faces, and so on down to the vertices, which are 0-dimensional. We let F_n be the number of n-dimensional 'faces' of the map. So for a polytope in 4-space, F_0 is the number of vertices, F_1 the number of edges, F_2 the number of faces, F_3 the number of solids, and F_4 the number of 4-dimensional regions – which for a regular polytope is 1.

The formula in two dimensions was

$$V-E+F = 1$$

or

$$F_0-F_1+F_2 = 1.$$

Bearing in mind how this formula was proved, by 'collapses', where changes in adjacent dimensions cancelled out, we are led to consider the expression

$$F_0 - F_1 + F_2 - F_3 + F_4$$

in 4-space. We can try this out on the regular polytopes, remembering that for these $F_4 = 1$. Using the table on page 203 we find that the expression takes the values

$$5 - 10 + 10 - 5 + 1 = 1$$
$$16 - 32 + 24 - 8 + 1 = 1$$
$$8 - 24 + 32 - 16 + 1 = 1$$
$$24 - 96 + 96 - 24 + 1 = 1$$
$$600 - 1200 + 720 - 120 + 1 = 1$$
$$120 - 720 + 1200 - 600 + 1 = 1,$$

which would be remarkable if it were a coincidence.

For a 3-dimensional map, the analogous expression would be

$$F_0 - F_1 + F_2 - F_3.$$

We try this out in something less regular (Figure 156).

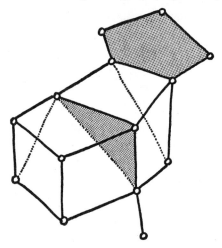

Figure 156

We have $F_0 = 14, F_1 = 22, F_2 = 11, F_3 = 2,$ and

$$14 - 22 + 11 - 2 = 1$$

which again suggests something more than mere coincidence.

For a map in n-dimensional space, we expect the equation

$$F_0 - F_1 + F_2 - \ldots \pm F_n = 1.$$

We can prove that this is the case without much difficulty: our collapsing technique is equal to the task. We can collapse simultaneously a vertex and an edge, or an edge and a face, or a face and a solid, . . . and in general an *m*-face and an $(m+1)$-face (Figure 157).

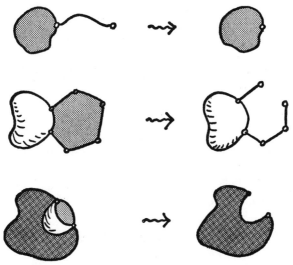

Figure 157

Each such collapse leaves the left-hand side of the equation unchanged; and eventually we are reduced to a single point, for which the value is 1. (To get the proof working properly it is necessary to do the collapses in the right order; but we have got the main idea.)

So in this case the method of proof, as well as the theorem, can be generalized.

The *n*-dimensional version of Euler's formula was first proved by Poincaré, and is therefore known as the *Euler–Poincaré* formula.

More Algebraic Topology

The idea of homotopy, and the fundamental group of Chapter 13, can be generalized to higher dimensions. Instead of paths obtained from a line segment, we use an *n*-dimensional hypercube. Instead of joining end to end we join face to face as in Figure 158. To obtain a group we look at hypercubes whose boundary is squashed up to a single point.

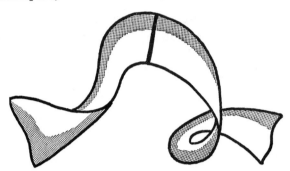

Figure 158

The notion of homotopy generalizes suitably, and we end up with a group, whose elements are homotopy classes of n-dimensional 'paths'. This is the n-th *homotopy group* $\pi_n(S)$ of the space S. The fundamental group $\pi(S)$ is now $\pi_1(S)$, the first of a whole series of algebraic invariants.

The higher homotopy groups can detect differences which π_1 misses. If we remove a spherical hole from a solid ball, to get a space S like the thickened-up peel of an orange, then $\pi_1(S)$ is trivial. Any loop can be slid over the hole and shrunk to a point. But if we put in a square, surrounding the hole, whose boundary is squashed up to a point (rather like putting a paper bag around the hole) then this square cannot be shrunk to a point inside S. So $\pi_2(S)$ is not trivial, and detects a hole which is missed by π_1.

If we know all the homotopy groups $\pi_1(S)$, $\pi_2(S)$, $\pi_3(S)$, ..., it might be hoped that we would know what S was, up to topological equivalence. Unfortunately this is not the case. However, Poincaré conjectured that a special case might be true: if S has the same sequence of πs as an n-sphere, then S is an n-sphere. For $n = 2$ this is essentially our assertion B of Chapter 12. For $n \geq 5$ it is also known to be true, and was proved by Smale.[1] But for $n = 3$ or 4, nobody knows. *

Thus the topology of higher dimensions can be easier than that of lower dimensions, which is a surprise. Indeed, it is part of topological folklore that the worst dimension is 4. Just what is so special about 4-space is an unsolved mystery.

*The Poincaré conjecture in four dimensions was proved by Michael Freedman in 1982. The three-dimensional version remains open. See Ian Stewart, *The Problems of Mathematics,* Oxford University Press, 1992. [1995 note.]

Chapter 15   Linear Algebra

A Problem

Early on in school algebra we are taught how to solve 'simultaneous equations' such as

$$x+2y = 6$$
$$3x-y = 4 \qquad (1)$$

by the following method (or a variant thereof): multiply the first equation by 3 to get

$$3x+6y = 18$$

and subtract the second equation, which gives

$$7y = 14$$

whence $y = 2$. Substitute this value in the first equation, which gives

$$x+4 = 6$$

and solve this for x, with the result $x = 2$.

Suppose instead we had started with the equations

$$x+2y = 6$$
$$3x+6y = 4. \qquad (2)$$

The method tells us to multiply the first by 3, giving

$$3x+6y = 18$$

and subtract the second, which yields

$$0 = 14.$$

We now attempt to solve this for y, and fail dismally. At school we were protected against such an event by a careful choice of questions. We were also protected from a related phenomenon, exemplified by these equations:

$$x+2y = 6$$
$$3x+6y = 18 \qquad (3)$$

where the standard procedure leads to the equation

$$0 = 0$$

We could just shake our heads, declare equations like (2) and (3) silly, and ignore the matter. But are we certain that we can always detect when something silly is going to occur?

With the particular equations given, an explanation of the behaviour is not hard to find. In equations (2) the two parts contradict each other, and there are no solutions. In (3) the second equation says the same as the first, so really all we have is a single equation connecting two unknowns. This does not mean that there are no solutions to (3): on the contrary there are many. For example $x = 2$, $y = 2$; $x = 4$, $y = 1$; $x = 6$, $y = 0$; $x = 1/2$, $y = 11/4$. On the other hand, we do not have absolute freedom: $x = 1$, $y = 1$ is *not* a solution. The full set of solutions can be found as follows: take any value for x, say $x = a$. For this choice of a we must take $y = \dfrac{6-a}{2}$. These, and only these, give solutions.

Thus there may be

$$\begin{cases} \text{a single solution,} \\ \text{no solutions,} \\ \text{infinitely many solutions,} \end{cases}$$

depending on the system of equations. In fact these are the only possibilities: one cannot find a system of simultaneous equations with exactly 2 real solutions, or exactly 3, or 4, or any finite number other than 0 and 1. (I won't prove this here, but it will be clear from the subsequent discussion that it is true.)

The same kind of behaviour occurs with more variables, and is far harder to detect. Faced with the equations

$$x+4y-2z+3t = 9$$
$$2x-y-z-t = 4$$
$$5x+7y+z-2t = 7$$
$$3x-2y-8z+5t = 21$$

one might be forgiven for not noticing immediately that twice the first equation plus three times the second minus the third gives

$$3x-2y-8z+5t = 23$$

contradicting the fourth equation. If we change 21 to 23 in the original system we get essentially 3 equations in 4 unknowns, which turn out to have infinitely many solutions.

It is not even true that if we have more unknowns than equations we can find solutions: the system

$$x+y+z+t = 1$$
$$2x+2y+2z+2t = 0$$

has no solutions.

Simultaneous equations, then, are not the tame things we are encouraged to believe them to be. Their behaviour is wild and (at first sight) unpredictable. If only it were true that all the simultaneous equations one actually encountered gave a single solution then we would be able to ignore the difficulties. Unfortunately, this is not so. Happily there are patterns concealed in the equations which allow us to resolve most of the problems.

A Geometric View

The conventional technique of plotting graphs of equations gives some explanation of what makes (1), (2), and (3) so different. In equations (1) the two parts correspond to lines, as shown in Figure 159: the unique solution is where the lines cross.

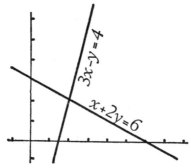

Figure 159

In equations (2) the lines are parallel (Figure 160) and never cross.

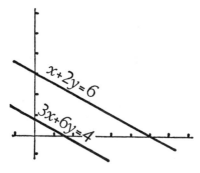

Figure 160

In equations (3) the lines *coincide*: thus all points on them correspond to solutions. (See Figure 161.)

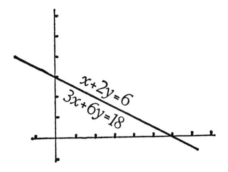

Figure 161

Clearly these are the only possibilities for two lines, which explains why we must have either 0, 1 or infinitely many solutions.

There is another way of looking at the equations geometrically which is more useful for a study of the general problem. What we do is this: we set up *two* sets of coordinates (x, y) and (X, Y). For each point (x, y) on the first graph we plot the point (X, Y) such that

$$x+2y = X$$
$$3x-y = Y.$$

Our original problem (1) now asks for us to find the point (x, y) which yields $(X, Y) = (6, 4)$.

What happens? Let's calculate (X, Y) for a few choices of (x, y).

(x, y)	(X, Y)
$(0, 0)$	$(0, 0)$
$(0, 1)$	$(2, -1)$
$(0, 2)$	$(4, -2)$
$(1, 0)$	$(1, 3)$
$(1, 1)$	$(3, 2)$
$(1, 2)$	$(5, 1)$
$(2, 0)$	$(2, 6)$
$(2, 1)$	$(4, 5)$
$(2, 2)$	$(6, 4)$

These results are illustrated in Figure 162.

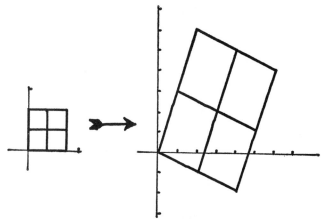

Figure 162

Clearly the *transformation* of (x, y) into (X, Y) turns squares in the (x, y)-plane into *parallelograms* in the (X, Y)-plane.

Our original equation is solved by looking at the last entry in the table: $x = 2$, $y = 2$ gives $X = 6$, $Y = 4$. This is fortuitous. But much more is clear from the picture. Consider *any* point (a, β) on the (X, Y)-plane. Obviously some point of the (x, y)-plane will end up at (a, β) because all we've done is stretch the plane a bit and rotate it. The parallelogram which contains (a, β) comes from some square: and some point in this square actually ends up at (a, β). For instance, take $(a, \beta) = (4\frac{1}{2}, 3)$ which lies in the middle of one of the parallelograms shown. The middle point of the corresponding square is $(1\frac{1}{2}, 1\frac{1}{2})$. And sure enough, the unique solution of

$$x + 2y = 4\frac{1}{2}$$
$$3x - y = 3$$

is $x = 1\frac{1}{2}, y = 1\frac{1}{2}$.

Further, it is clear that this solution *must* be unique, because the way squares change to parallelograms will not allow two distinct points in the (x, y)-plane to end up in the same place. The transformation does not introduce folds.

If however we plot what happens for equations (2) (and, as it happens, (3)), we must look at

$$x+2y = X$$
$$3x+6y = Y.$$

It is not hard to see that the only values of (X, Y) which arise lie on a *line*: namely the line $Y = 3X$ (Figure 163).

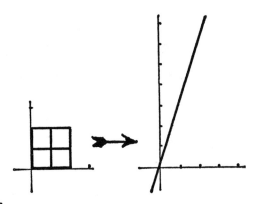

Figure 163

The relevant transformation squashes the whole (x, y)-plane up into a single line. In equation (2) we have to solve

$$X = 6, Y = 4.$$

But (6, 4) does not lie on the line. So no solution is possible: *no* (x, y) can end up off the line. On the other hand, for equations (3) we want to solve

$$X = 6, Y = 18$$

and now (6, 18) *is* on the line: furthermore the squashing gives infinitely many (x, y) which end up at (6, 18).

Further, all the possible values of (x, y) form a line in the (x, y)-plane, given by

$$x +2y = 6.$$

Thus the different phenomena which occur depend upon the geometric properties of the transformation T such that

$$T(x, y) = (X, Y) = (x+2y, 3x-y)$$

and the transformation S such that

$$S(x, y) = (X, Y) = (x+2y, 3x+6y).$$

In order to study the general equation

$$ax+by = X$$
$$cx+dy = Y$$

we should look at the transformation

$$U(x, y) = (ax+by, cx+dy).$$

For equations in 3 unknowns

$$ax+by+cz = X$$
$$dx+ey+fz = Y$$
$$gx+hy+kz = Z$$

we want the transformation

$$V(x\ y, z) = (ax+by+cz, dx+ey+fz, gx+hy+kz).$$

These are known as *linear* transformations: their study is called *linear algebra*.

A Hint of a Pattern

We can reformulate equations (1) set-theoretically by using the linear transformation T above. They become this: is the point $(6, 4)$ an element of the *range* of T? Recall that the range of T is the set of values $T(x, y)$ which T takes: then $(6, 4)$ is in it if and only if there exist x and y such that $(6, 4) = T(x, y) = (x+2y, 3x-y)$; which is the same as equations (1).

The same goes for the other two equations: they ask whether $(6, 4)$ or $(6, 18)$ are in the range of the linear transformation S.

From the way that T takes squares to parallelograms it is geometrically obvious that the range of T is the whole plane. As we found, the range of S is a single line.

If we wish to study general simultaneous equations, we should therefore try to find out about the ranges of linear transformations. So far we have found planes and lines. Are these all?

Not quite. The trivial simultaneous equation

$$0x+0y = X$$
$$0x+0y = Y$$

corresponds to the linear transformation F with $F(x, y) = (0, 0)$. The range of F is a single point, $\{(0, 0)\}$. (The curly brackets are just me being pedantic about set theory: the range is a set.)

However, this exhausts the possibilities for 2 equations in 2 unknowns. The range is either a plane, a line, or a point. Of course if it is a plane it must be the whole of $\mathbf{R}^2$.

When the range is a plane, solutions always exist and are unique. When the range is a line, solutions may or may not exist; they exist exactly for (X, Y) on this line; and for a fixed (X, Y) the possible solutions themselves form a line. When the range is a point, solutions exist only for $(X, Y) = (0, 0)$ and then we have a whole *plane* of solutions: namely anything in $\mathbf{R}^2$.

Let us introduce the term 'solution space' for the set of solutions, when such exist. Then we have the following possibilities:

range	solution space
plane	point
line	line
point	plane

For 3 equations in 3 unknowns we would expect the ranges and solution spaces to be either points, lines, planes, or 'solids' (i.e. the whole of $\mathbf{R}^3$). This is indeed the case. Furthermore we have:

range	solution space
solid	point
plane	line
line	plane
point	solid

In other words, the smaller the range the bigger the solution space: but on the other hand the smaller the range the less chance that solutions exist at all.

The same sort of thing happens in general. If we look at $\mathbf{R}^n$ it can be proved that the dimension of the range plus the dimension of the solution space is n. Thus a linear transformation on $\mathbf{R}^7$ with range of dimension 3 would have a 4-dimensional solution space.

Of course I have not yet defined *dimension*. This is where the subject of linear algebra really starts, but the details are best found in a proper textbook.[1] However, it should be clear that the apparent wildness of simultaneous equations can be organized into a clear pattern of behaviour.

Matrices

There is a useful notation for linear transformations, invented by Cayley. If $T(x, y) = (X, Y)$ where

$$ax + by = X$$
$$cx + dy = Y \qquad (4)$$

we pick out the coefficients and write them in a square array, thus:

$$\begin{pmatrix} a & b \\ c & d \end{pmatrix}.$$

Such an expression is called a *matrix*: the matrix of T. If we know this matrix then we know T, provided we know which variables x, y, X, Y we are using. We can bring them into the picture too by introducing *column vectors*

$$\begin{pmatrix} x \\ y \end{pmatrix} \qquad \begin{pmatrix} X \\ Y \end{pmatrix}$$

and write (4) compactly as

$$\begin{pmatrix} a & b \\ c & d \end{pmatrix}\begin{pmatrix} x \\ y \end{pmatrix} = \begin{pmatrix} X \\ Y \end{pmatrix} \qquad (5)$$

where by definition the 'product' on the left is the column vector

$$\begin{pmatrix} ax + by \\ cx + dy \end{pmatrix}$$

and two column vectors are equal if and only if their entries are equal.

This notation can easily be extended to 3 or more unknowns. The general system of 3 equations in 3 unknowns takes the form

$$\begin{pmatrix} a & b & c \\ d & e & f \\ g & h & k \end{pmatrix}\begin{pmatrix} x \\ y \\ z \end{pmatrix} = \begin{pmatrix} X \\ Y \\ Z \end{pmatrix}.$$

Often we have to deal with several transformations one after another. We might have yet more variables $\mathbf{X}$ and $\mathbf{Y}$ and a transformation U such that $U(X, Y) = (\mathbf{X}, \mathbf{Y})$ where

$$AX + BY = \mathbf{X}$$
$$CX + DY = \mathbf{Y} \qquad (6)$$

or, in matrix notation,

$$\begin{pmatrix} A & B \\ C & D \end{pmatrix}\begin{pmatrix} X \\ Y \end{pmatrix} = \begin{pmatrix} \mathbf{X} \\ \mathbf{Y} \end{pmatrix} \tag{7}$$

We have already defined a product of transformations, which gives

$$UT(x, y) = U(X, Y) = (\mathbf{X}, \mathbf{Y}).$$

Now UT, it turns out, can also be represented by a matrix. From (6) and (4) we get

$$\begin{aligned}
\mathbf{X} &= AX + BY \\
&= A(ax + by) + B(cx + dy) \\
&= (Aa + Bc)x + (Ab + Bd)y, \\
\mathbf{Y} &= CX + DY \\
&= C(ax + by) + D(cx + dy) \\
&= (Ca + Dc)x + (Cb + Dd)y.
\end{aligned}$$

Picking out the coefficients we could write

$$\begin{pmatrix} Aa + Bc & Ab + Bd \\ Ca + Dc & Cb + Dd \end{pmatrix}\begin{pmatrix} x \\ y \end{pmatrix} = \begin{pmatrix} \mathbf{X} \\ \mathbf{Y} \end{pmatrix}$$

which gives the matrix of UT as

$$\begin{pmatrix} Aa + Bc & Ab + Bd \\ Ca + Dc & Cb + Dd \end{pmatrix}.$$

On the other hand, we could work purely formally to obtain from (5) and (7) the equation

$$\begin{pmatrix} A & B \\ C & D \end{pmatrix}\begin{pmatrix} a & b \\ c & d \end{pmatrix}\begin{pmatrix} x \\ y \end{pmatrix} = \begin{pmatrix} \mathbf{X} \\ \mathbf{Y} \end{pmatrix}$$

which *suggests* that we might get a nice kind of algebra if we define a product for matrices by

$$\begin{pmatrix} A & B \\ C & D \end{pmatrix}\begin{pmatrix} a & b \\ c & d \end{pmatrix} = \begin{pmatrix} Aa + Bc & Ab + Bd \\ Ca + Dc & Cb + Dd \end{pmatrix}.$$

For example, in Chapter 2 we had transformations G and H where $G(x, y) = (x, -y)$ and $H(x, y) = (y, -x)$. If $(X, Y) = H(x, y)$ then we have

$$\begin{aligned}
X &= y & = 0.x + 1.y \\
Y &= -x & = (-1).x + 0.y
\end{aligned}$$

so the matrix of H is

$$\begin{pmatrix} 0 & 1 \\ -1 & 0 \end{pmatrix}.$$

If $G(X, Y) = (\mathbf{X}, \mathbf{Y})$ then

$$\mathbf{X} = X = 1.X + 0.Y$$
$$\mathbf{Y} = -Y = 0.X + (-1).Y$$

and the matrix of G is

$$\begin{pmatrix} 1 & 0 \\ 0 & -1 \end{pmatrix}.$$

By the formula the matrix of GH should be

$$\begin{pmatrix} 1 & 0 \\ 0 & -1 \end{pmatrix}\begin{pmatrix} 0 & 1 \\ -1 & 0 \end{pmatrix}$$

which is

$$\begin{pmatrix} 1.0+0.(-1) & 1.1+0.0 \\ 0.0+(-1)(-1) & 0.1+(-1).0 \end{pmatrix}$$

which simplifies to

$$\begin{pmatrix} 0 & 1 \\ 1 & 0 \end{pmatrix}.$$

Now we found that $GH(x, y) = (y, x)$, so that

$$\mathbf{X} = 0.x + 1.y$$
$$\mathbf{Y} = 1.x + 0.y$$

which checks.

In fact we do get a nice kind of algebra by making this definition: one which allows us to calculate with linear transformations. I don't want to go into details because there is already an excellent treatment by Sawyer.[2]

However, let me do one more calculation which shows how the use of matrices can give rise to results in trigonometry. In Chapter 2 I gave a formula for the transformation 'rotate through angle θ'. In matrix form it is

$$\begin{pmatrix} \cos\theta & -\sin\theta \\ \sin\theta & \cos\theta \end{pmatrix}$$

Thus the product of a rotation through θ and a rotation through ϕ has matrix

$$\begin{pmatrix} \cos\phi & -\sin\phi \\ \sin\phi & \cos\phi \end{pmatrix}\begin{pmatrix} \cos\theta & -\sin\theta \\ \sin\theta & \cos\theta \end{pmatrix}$$

which works out as

$$\begin{pmatrix} \cos\phi\cos\theta-\sin\phi\sin\theta & -\cos\phi\sin\theta-\sin\theta\cos\phi \\ \sin\phi\cos\theta+\cos\phi\sin\theta & -\sin\phi\sin\theta+\cos\phi\cos\theta \end{pmatrix}.$$

But of course it should represent a rotation through $(\phi+\theta)$, with matrix

$$\begin{pmatrix} \cos(\phi+\theta) & -\sin(\phi+\theta) \\ \sin(\phi+\theta) & \cos(\phi+\theta) \end{pmatrix}.$$

Comparing the two expressions gives the equations

$$\cos(\phi+\theta) = \cos\phi\cos\theta-\sin\phi\sin\theta$$
$$\sin(\phi+\theta) = \sin\phi\cos\theta+\cos\phi\sin\theta$$

which are the 'addition formulae' of trigonometry.

An Abstract Formulation

Nowadays the study of linear transformations is a part of abstract algebra. This arises from attempts to avoid using coordinates in the theory.

Given two points (p, q) and (r, s) of $\mathbf{R}^2$ we can define their sum to be

$$(p, q)+(r, s) = (p+q, r+s).$$

If we have a real number a we can also define a product

$$a(p, q) = (ap, aq).$$

By means of these operations we can characterize linear transformations: they are exactly those functions $T:\mathbf{R}^2 \to \mathbf{R}^2$ such that for all p, q, r, s and a we have

$$T((p, q)+(r, s)) = T(p, q)+T(r, s)$$
$$T(a(p, q)) = aT(p, q).$$

(You can check this if you wish.) The first equation is very similar to that obeyed by an isomorphism in the sense of group theory, which suggests that an abstract approach along group-theoretic lines might be illuminating. By seeing what properties hold for addition and multiplication by real numbers in $\mathbf{R}^2$, and analogous operations in $\mathbf{R}^3 \mathbf{R}^4, \mathbf{R}^5, \ldots$, mathematicians evolved the following formulation.

A *vector space* over **R** is a set V with two operations, called *addition* and *scalar multiplication*. If u and v are elements of V and a is a real number then the results of these operations are denoted by

$$u+v \qquad au$$

respectively: $u+v$ and au are elements of V.
The following axioms must hold:

(1) V is a commutative group under addition, with identity element 0.

(2) $a0 = 0$ for all $a \in \mathbf{R}$.

(3) $0v = 0$ for all $v \in V$.

(4) $1v = v$ for all $v \in V$.

(5) $(a+\beta)v = av+\beta v$ for all a, $\beta \in \mathbf{R}$, $v \in V$.

(6) $a(v+w) = av+aw$ for all $a \in \mathbf{R}$, v, $w \in V$.

(7) $a\beta v = a(\beta v)$ for all a, $\beta \in \mathbf{R}$, $v \in V$.

There are many examples of vector spaces. The standard ones are $\mathbf{R}$, $\mathbf{R}^2$, $\mathbf{R}^3$, ... but these are not the only ones. The polynomial ring $\mathbf{R}[x]$ in one indeterminate is a vector space: so are $\mathbf{R}[x, y]$, $\mathbf{R}[x, y, z]$, ... These have *infinite* dimension. Vector spaces arise in the solution of differential equations, in certain parts of group theory, and in modern formulations of the calculus.

A *linear transformation* is now defined as a function $T: V \to W$ where V and W are arbitrary vector spaces, with the properties

$$T(u+v) = T(u)+T(v)$$
$$T(au) = aT(u)$$

for all u, $v \in V$, $a \in \mathbf{R}$.

In this abstract formulation one can prove all the desired theorems about linear transformations. Because no particular choice of coordinates is made, the proofs are very clean and direct.

However, to perform calculations in particular cases one uses matrix notation.

A proper understanding of linear algebra requires a synthesis of three points of view:

(i) the underlying geometrical motivation,

(ii) the abstract algebraic formulation,

(iii) the matrix-theoretic technique.

This makes matters hard for the student, to begin with; which is probably why most textbooks concentrate on one of the three viewpoints. But in the long run this bias will cause more problems than it solves: the sight of a student struggling with enormous matrices when a little geometrical insight would solve the whole problem in two lines is not an inspiring one.

Chapter 16   Real Analysis

'There is in mathematics hardly a single infinite series of which the sum is determined in a rigorous way' –
N. H. Abel, in a letter of 1826

The three cornerstones of modern mathematics are algebra, topology, and *analysis.* (Mathematical logic is more like the mortar which holds the bricks together.) I have treated the first two at some length; so it is only fair to say something about the third.

It is an unfortunate fact of life that analysis cannot be discussed to any great depth without introducing many technical concepts. A naive approach to analysis runs into insurmountable obstacles, as any history of mathematics will demonstrate.

Analysis might be described as the study of *infinite processes,* such as infinite series, limits, differentiation, and integration. It is the spectre of the infinite looming above us which causes the difficulties.

Infinite Addition

An *infinite series* is an expression such as
$$1+\tfrac{1}{2}+\tfrac{1}{4}+\tfrac{1}{8}+\dots. \qquad (1)$$
Its essence is the '. . .', which seems to ask us to continue adding terms infinitely often. It is well to view such an expression with scepticism, for it appears to be asking us to carry out an impossible process: there is no man alive, no computer, however fast, that can do infinitely many additions in a finite time. One is reminded of paradoxical questions about light switches that are switched on after one second, off half a second later, on after a quarter of a second, off after an eighth, . . . ; after two seconds, is the switch on or off?

Thus on the face of it we have no guarantee that the expression (1) means anything at all – an observation which escaped the attention of almost everybody who worked on the subject in the eighteenth century. It then seemed that any kind of combination of mathematical symbols was mathematically meaningful; a naive notion of which mathematicians were disabused, painfully, as time went on.

However, if (1) does mean anything, then the best guess as to what it might mean is surely the number 2. For we have

$$1+\frac{1}{2} \qquad\qquad =\frac{3}{2}$$

$$1+\frac{1}{2}+\frac{1}{4} \qquad\qquad =\frac{7}{4}$$

$$1+\frac{1}{2}+\frac{1}{4}+\frac{1}{8} \qquad\qquad =\frac{15}{8}$$

$$\cdots$$

$$1+\frac{1}{2}+\frac{1}{4}+\frac{1}{8}+\ldots+\frac{1}{2^n} = 2-\frac{1}{2^n}.$$

If we think that the sum should be 2, then the error we commit by stopping after $n+1$ terms is $1/2^n$. As n gets larger 2^n gets very much larger, and $1/2^n$ becomes rapidly smaller. In fact, by taking n large enough, we can make $1/2^n$ as small as we please.

In the eighteenth century matters would have been phrased thus: in the expression for the sum of $n+1$ terms put $n = \infty$. Then the left-hand side is the sum of $\infty+1$ terms, but $\infty+1 = \infty$ so this is the series (1). The right-hand side, on the other hand, is

$$2- \frac{1}{2^\infty} = 2-\frac{1}{\infty} = 2-0 = 2.$$

This proves that the sum must be 2.

Actually it does nothing of the kind, for at least three reasons. Firstly, we must assume that (1) has a meaning. Secondly, we must assume that infinite sums can be subjected to algebraic operations as if they were finite sums. Thirdly, the use of ∞ as a symbol for 'infinity' assumes that ∞ behaves like a number: is this assumption justifiable?

A blind manipulation of infinite series leads to all sorts of paradoxes, several of them quite charming, all of them mathematically disastrous.

Thus let

$$S = 1-1+1-1+1-1+\ldots.$$

Then

$$
\begin{aligned}
S &= (1-1)+(1-1)+(1-1)+\ldots \\
&= 0+0+0+\ldots \\
&= 0.
\end{aligned}
$$

Equally,

$$
\begin{aligned}
S &= 1-(1-1)-(1-1)-(1-1)-\ldots \\
&= 1-(0+0+0+\ldots) \\
&= 1-0 \\
&= 1.
\end{aligned}
$$

Or again,

$$
\begin{aligned}
1-S &= 1-(1-1+1-1+1-1+\ldots) \\
&= 1-1+1-1+1-1+\ldots \\
&= S
\end{aligned}
$$

and we can solve for S, obtaining $S = 1/2$.

One hopeful spirit went as far as to suggest that the equality of 0 and 1 (expressed via S) was symbolic of creation from nothing. Not only does this justify the uncritical manipulation of infinite processes; it also gives a mathematical proof of the existence of God!

The early days of analysis were bedevilled by a feeling that somehow all three values for S are 'correct'. Mathematics had not yet learnt to cut its losses. It was gradually realized that infinite processes are not of themselves meaningful: they must be *given* a meaning. Once this has been done, restrictions may be imposed on the expressions which occur in the processes. Further, one can no longer assume that the usual laws are obeyed, though with luck something may be salvaged.

What is a Limit?

Let's take a closer look at that troublesome series S. The sums to 1, 2, 3, 4, . . . terms are

$$
\begin{aligned}
1 &= 1 \\
1-1 &= 0 \\
1-1+1 &= 1 \\
1-1+1-1 &= 0 \\
1-1+1-1+1 &= 1
\end{aligned}
$$

which take alternately the values 0 and 1. As n gets larger and larger these do not settle down to some kind of 'limit'; they just hop happily from 0 to 1 and back again.

If we think the sum is 1 we commit an error of 1 at all even terms; if 0 then we are in error by 1 at all odd terms. In fact the best bet is 1/2, because this has the virtue of minimizing the error at all stages!

A general series will look like

$$
a_1 + a_2 + a_3 + \ldots
$$

where the as are real numbers. The 'approximate' sums are

$$
\begin{aligned}
b_1 &= a_1 \\
b_2 &= a_1 + a_2 \\
b_3 &= a_1 + a_2 + a_3 \\
b_4 &= a_1 + a_2 + a_3 + a_4 \\
&\ldots
\end{aligned}
$$

If the values of b_n settle down to some 'limit' as n gets very large, we could *define* the value of the series to be this limit. What do we mean by 'limit'?

A first hope is to look at the *error* we commit in stopping after n terms: if a limit exists this error should become very small. But the error is

$$
a_{n+1} + a_{n+2} + a_{n+3} + \ldots
$$

which is another *infinite* series. As far as we can see this won't help us much.

We must therefore concentrate on the sequence of approximate sums

$$
b_1, b_2, b_3, b_4, \ldots
$$

and see if we can give a meaning to the 'limit' of this sequence.

Example (1) should help, because we do expect it to have a sum, namely 2. For it, the approximate sums are

$$
b_n = 2 - \frac{1}{2^{n-1}}.
$$

The difference $b_n - 2$ can be made as small as we please by taking n large enough. Thus to make

$$-\frac{1}{1\,000\,000} \leq b_n - 2 \leq +\frac{1}{1\,000\,000}$$

we only need make

$$1/2^{n-1} \leq 1/1\,000\,000$$

which works for $n \geq 21$. To make

$$-\frac{1}{1\,000\,000\,000\,000} \leq b_n - 2 \leq +\frac{1}{1\,000\,000\,000\,000}$$

we need $n \geq 41$. And so on.

This, and similar examples, leads to a definition. The sequence b_n is said to *tend* to the *limit* l if the difference

$$b_n - l$$

can be made as small as we please by taking n sufficiently large.[1]

A sequence which does tend to a limit is said to be *convergent*. (The limit l must be a *real* number: we do not at this stage talk about ∞.)

Having defined 'limit' for the b_n we can give a meaning to the infinite series

$$a_1 + a_2 + a_3 + a_4 + \dots.$$

It is the limit l of the sequence b_n of approximating sums, *provided that limit exists*. If it does, we say that the series is *convergent*. It need not always exist, as the troublesome series S demonstrates.

We can now talk about the sum of an infinite series, *but only if we can prove it convergent*. (There are other, less natural, definitions which allow us to assign a sum to series not convergent in our sense – in fact the troublesome S turns out to be well-behaved in some of these theories, and has sum $1/2$ – but we won't discuss these.)

Once we can talk about sums, we can ask about the laws of algebra. Can we put brackets in wherever we want, or rearrange terms?

Even for convergent series, this is not always possible. The series

$$K = 1 - \tfrac{1}{2} + \tfrac{1}{3} - \tfrac{1}{4} + \tfrac{1}{5} - \tfrac{1}{6} \dots$$

can be proved convergent: in fact its sum is $\log_e 2$ which is about 0·69.

What then, is wrong with the following argument?[2]

$$2K = 2 - \tfrac{2}{2} + \tfrac{2}{3} - \tfrac{2}{4} + \tfrac{2}{5} - \tfrac{2}{6} + \tfrac{2}{7} - \tfrac{2}{8} + \tfrac{2}{9} - \tfrac{2}{10} + \dots$$
$$= 2 - 1 + \tfrac{2}{3} - \tfrac{1}{2} + \tfrac{2}{5} - \tfrac{1}{3} + \tfrac{2}{7} - \tfrac{1}{4} + \tfrac{2}{9} - \tfrac{1}{5} + \dots$$
$$= (2-1) - \tfrac{1}{2} + (\tfrac{2}{3} - \tfrac{1}{3}) - \tfrac{1}{4} + (\tfrac{2}{5} - \tfrac{1}{5}) - \dots$$
$$= 1 - \tfrac{1}{2} + \tfrac{1}{3} - \tfrac{1}{4} + \tfrac{1}{5} - \dots$$
$$= K.$$

Therefore $1 \cdot 38 = 0 \cdot 69$.

The Completeness Axiom

Our definition of convergence has one unsatisfactory aspect: before we can prove that a series converges we have to guess the limit *l* towards which it converges. Thus it is relatively easy to prove that (1) converges once we have guessed that the limit should be 2. As yet we have no way of testing for convergence other than guessing a limit.

Progress on this depends on taking another look at the 'error' terms

$$a_{n+1} + a_{n+2} + a_{n+3} + \dots$$

which we prematurely dismissed as useless. For a convergent series, these errors are 'small'. Can we make this idea precise, and use it as a basis for a convergence criterion?

Let's try to approximate the error. (This looks doomed to failure, because we must then take into account the error in *this*, but we may as well see.) We get now

$$a_{n+1}$$
$$a_{n+1} + a_{n+2}$$
$$a_{n+1} + a_{n+2} + a_{n+3}$$
$$\dots$$
$$a_{n+1} + \dots + a_{n+m}.$$

Suppose that these are *all* small. In fact, suppose that there exists some small positive number *k* such that

$$-k \le a_{n+1} + \dots + a_{n+m} \le k$$

for *every* m. Then it is reasonable to say that the whole error term is 'small', and that its size is $\le k$.

In other words, a convergent series should have the following property. Pick any $k > 0$. Then we can find an integer n (depending on k) such that for any m the 'approximate error'

$$a_{n+1}+\ldots+a_{n+m}$$

is smaller than k.

And conversely if we can do this, then the errors get arbitrarily small: we would expect the series to converge.

Now the nice thing about the above idea is that it doesn't involve any infinite sums, and it doesn't involve guessing a limit. It talks only about finite sums of terms from the series. In return for this we have to handle a statement whose logic is fairly complicated, but no matter.

It is here that the real numbers come in. If we thought that all numbers were rational we could define a limit of a sequence of rational numbers to be a rational number l with the property that the members of the sequence get arbitrarily close to l. We could also make the above analysis of error terms.

Consider now the decimal expansion of $\sqrt{2}$:

$$\sqrt{2} = 1\cdot414213\ldots.$$

We can view this as an infinite series

$$1+\frac{4}{10}+\frac{1}{100}+\frac{4}{1\,000}+\frac{2}{10\,000}+\frac{1}{100\,000}+\frac{3}{1\,000\,000}+\ldots.$$

Now the approximate error in stopping after n terms is at most

$$0\cdot\underbrace{00\ldots0}_{n}\underbrace{999\ldots9}_{m}$$

and however large we take m this is less than $1/10^n$. For large enough n, $1/10^n$ becomes arbitrarily small. We would therefore expect the series to converge.

On the other hand, if it does converge, it must converge to $\sqrt{2}$, which is not a rational number.

Nonetheless it is a series all of whose terms are rational; and in our ignorance we expect it to have a rational sum: but it doesn't. This is an awkward phenomenon. Our intuitive feeling is that it is caused by the absence from the rationals of certain numbers like $\sqrt{2}$. We get the real numbers by filling in the holes.

For logical precision, we add an extra axiom to the axioms for rationals, known as the *completeness axiom*. It ensures the existence of a *real* number which is the limit of a given sequence for which the error terms become arbitrarily small.

Continuity

In the chapter on topology we came across the concept of a continuous function. The same concept is of fundamental importance in analysis.

If we draw the graph of a function such as

$$f(x) = 1 - 2x - x^2$$

(Figure 164) we see that there are no 'jumps' in the graph.

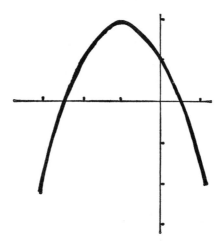

Figure 164

On the other hand, the graph of first-class postage rates for letters (at the time of writing) has a number of jumps (Figure 165).

The first function is continuous, the second discontinuous.

In the early days of analysis it was thought that any function defined by a nice formula must be continuous. This is a pious but vain hope: the function

$$g(x) = x + \sqrt{(x-1)(x-2)}$$

has the graph of Figure 166.

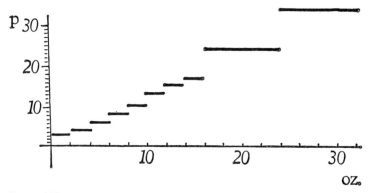

Figure 165

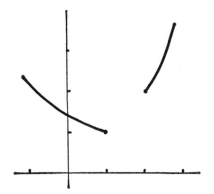

Figure 166

One must therefore be more careful. Euler tried to define a continuous function as a 'curve described by freely leading the hand' but this was not very helpful. Cauchy at first defined it as 'a function for which an infinitesimal change in the variable produces an infinitesimal change in the value'. This is nice if you know what an infinitesimal is, but nobody did: naive attempts to handle it suffered from the same rash of paradoxes that incapacitated the infinite.

The definition in current use is based upon the idea 'no jumps'.

As with the housemaid's baby, so with jumps: a small one is just as unhappy an occurrence as a big one. This means that we allow ourselves to use a microscope, so to speak, when looking for jumps. Under a microscope a jump will look something like Figure 167.

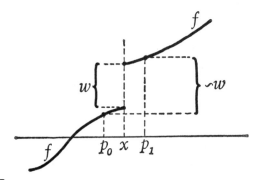

Figure 167

The jump has a definite width w. So if we take p_0 little to the left of x and p_1 a little to the right then the values $f(p_0)$ and $f(p_1)$ will differ by about $w(\sim w)$. However, if we take p_0 and p_1 too far away we can't expect to say anything much about $f(p_0)$ and $f(p_1)$.

The standard definition of continuity of a function f defined on the set of real numbers is designed to eliminate jumps. We say that f is *continuous at a point* x if by choosing p_0 and p_1 sufficiently close to x we can make $f(p_0)$ and $f(p_1)$ as close together as we please.[3] Then we say that f is *continuous* if it is continuous at all points x.

The advantage of this definition over Euler's is that we can *prove* certain functions to be continuous. For instance, take
$$f(x) = x^2.$$
To prove this continuous at 0 we note that if p_0 is taken between $-k$ and 0 and p_1 between 0 and k, for a positive number k, then
$$-2k^2 \leq p_0{}^2 - p_1{}^2 \leq 2k^2.$$
The size of the difference is $2k^2$, and by choosing k small enough we can make this as small as we please. To make $2k^2 \leq 1/1\,000\,000$ we need only take $k < 1/10\,000$, and so on. To prove f continuous we must do the same sort of thing at all points x, instead of just 0: the algebra is more complicated, but quite easy.

Functions can be continuous at some points and discontinuous at others. The postal-charge function is discontinuous at 2, 4, 6,

8, 10, 12, 14, 16, 24, and 32 oz., but continuous everywhere else in the range 0 to 32 oz.

However, some functions can behave in a very funny way, which shows that although we may have a good definition it might not be exactly the one we intended. The function h such that

$$h(x) = \begin{cases} 0 & \text{if } x \text{ is irrational} \\ 1/q & \text{if } x = p/q \text{ is rational} \end{cases}$$

is continuous at all irrational points but discontinuous at all rational points. Even more curiously, it is not possible to find a function which is continuous at all rational points but discontinuous at all irrational points!

Of course this is a very funny function.

The definition of continuity we have given turns out to be completely adequate for setting up all the machinery of analysis, despite its peculiarities. However, there may well be a better way of proceeding. Recently the idea of an 'infinitesimal' has been made respectable by using a rather complicated construction from mathematical logic. Cauchy's definition of continuity can be made rigorous. However, I would not advocate teaching 'nonstandard analysis', as it is called, to a student: the logic of the subject is exceedingly subtle.

In fact no one has yet found a very satisfactory way of introducing rigorous analysis, particularly at school level.

Proving Theorems in Analysis

A puzzle which was current not long ago concerns a man walking up a mountain. At exactly 9 o'clock on Monday morning he sets off up the mountain, arriving at 6 o'clock in the evening at a mountain hut, where he stays the night. At 9 o'clock the next morning he descends by the same route, arriving at the point from which he set out at 6 o'clock in the evening. *Prove that at some time he is at the same place on both days.*

A moment's thought might suggest an answer: on the second day imagine a ghost walking up the mountain, copying exactly what the man did on the first day. Since the ghost is going *up* and the man *down*, they must meet: this will provide the time of day asked for by the puzzle.

Hidden in all this is a theorem in analysis. For we have assumed that the man's progress is continuous. If, by some technological miracle the man could jump from one part of the mountain to another without passing through the part in between, he might be able to avoid the ghost.

We could draw a graph of the man's progress on the two days (Figure 168), in which case the idea of the proof stands out: *the two curves must cross.*

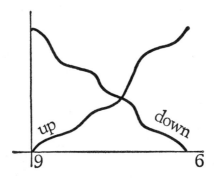

Figure 168

For discontinuous curves, this need not happen (Figure 169).

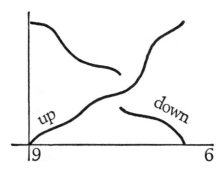

Figure 169

Now in analysis we cannot argue from a picture, because the picture may lie. We have to work logically from the definition.

(Of course, we should bear the picture in mind!) The theorem we want should say something like this: suppose we have two continuous functions f and g defined on the real line, such that at two points a and b we have

$$f(a) < g(a) \qquad f(b) > g(b).$$

Then at some point c between a and b we must have

$$f(c) = g(c).$$

(See Figure 170).

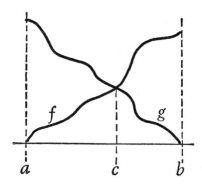

Figure 170

We could prove the theorem like this. Divide the interval between a and b into 10 parts. For some of these f stays smaller than g (Figure 171). Take the first one where f becomes greater than g,

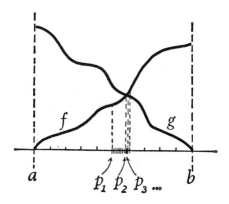

Figure 171

and divide it up into 10 parts. Take the first of these, divide it up into 10 parts . . . and so on. The end-points of the intervals will form a sequence

$$p_1, p_2, p_3, \ldots$$

and an application of the completeness axiom shows that this sequence converges to a real number p between a and b. A little more work, using the definition of continuity, shows that $f(p) = g(p)$.

If in Figure 171 we had $a = 0, b = 1$, then we would find that

$$p_1 = 0 \cdot 5$$
$$p_2 = 0 \cdot 58$$
$$p_3 = 0 \cdot 583$$
$$\ldots$$

and it is clear that the procedure leads to an infinite decimal

$$p = 0 \cdot 583 \ldots .$$

What the completeness axiom does, in fact, is to make infinite decimals respectable. And of course my reason for dividing into 10 parts at each stage was to fit in with decimal notation. It would have worked just as well if I had divided into 2 parts, or 19, or 1066.

The use of the completeness axiom in this theorem is essential, because the theorem is not true over the rational numbers. The function

$$f(x) = 1 - 2x - x^2$$

is a continuous function on the rationals. Further $f(0) = 1$, $f(1) = -2$. If the theorem were true for rationals there should be a rational number p between 0 and 1 such that

$$1 - 2p - p^2 = 0.$$

But this makes $p = \sqrt{2} - 1$, which is *not* rational. So we can't escape without using completeness somewhere.

We could, of course, dismiss the rigorous proof as being superfluous: if a theorem is geometrically obvious why prove it? This was exactly the attitude taken in the eighteenth century. The result, in the nineteenth century, was chaos and confusion: for intuition, unsupported by logic, habitually assumes that everything is much nicer behaved than it really is.

Good ideas in mathematics should never be ignored just because a rigorous foundation is lacking. But if they are allowed to develop too far without finding an underlying rationale they usually lead to trouble.

Chapter 17   The Theory of Probability

'Statistics is a branch of theology' –
A Cambridge research fellow

Probability theory has its origins in questions about gambling. In a game of cards, or dice, when do I have the best chance of winning? What are the odds?[1]

Because games are usually finite, the methods needed to handle such questions are *combinatorial*, that is, based on counting arguments. For example, to find the chance of throwing three consecutive heads with a coin one lists the possibilities

$$HHH \quad HHT \quad HTH \quad HTT$$
$$THH \quad THT \quad TTH \quad TTT$$

which are 8 in number. Exactly 1 is favourable, so the probability is 1/8.

This of course makes the assumption that throws of H or T are equally likely. Now we can't define 'equally likely' by saying 'probability 1/2' until we have defined what we mean by 'probability 1/2': and we can't do that without defining 'equally likely'. Or at least, so it seems.

If we try to get round it by doing experiments, we run into another difficulty. If H and T are equally likely, then in a long series of throws we would expect to have approximately equal numbers of H and T. Not exactly equal, of course: they couldn't possibly be equal in an odd number of throws anyway, and in an even number of throws there would probably be a small discrepancy. Toss a coin 20 times and see if you get exactly 10 heads. (If you do, try several more times and see how often it happens!)

What we would hope is that 'in the limit' the ratio of the number of Hs to the number of Ts should 'tend to' 1/2. The trouble is that this 'limit' is not a limit in the usual sense of analysis. It is conceivable that we might throw a sequence consisting entirely of Hs with a *fair* coin. It is, of course, unlikely. But to set up an idea of 'limit' which takes account of this possibility involves making precise what we mean by 'unlikely', which seems to require a definition of 'probability' again!

It wasn't until the 1930s that these difficulties were circumvented. This was achieved by developing an *axiomatic* probability theory. By divorcing the mathematics from its applications one can develop the mathematics without any logical qualms: *then* it can be tested experimentally to see if it fits the facts. Axiomatic probability theory succeeds for the same reason that axiomatic geometry succeeds.

Combinatorial Probability

For the moment assume that we know what 'equally likely' means. Then a rough working definition of the *probability* $p(E)$ of an *event E* is

$$p(E) = \frac{\text{number of ways in which } E \text{ can occur}}{\text{total number of possible occurrences}}$$

(provided all occurrences are equally likely).

Thus there are 36 ways of throwing 2 dice; and 5 of these give a total of 6 (namely $1+5, 2+4, 3+3, 4+2, 5+1$).

Therefore the probability that the total is 6 is

$$\frac{\text{number of ways of throwing 6}}{36}$$

which is 5/36.

Since the numbers involved are positive, and since the number of ways E can occur is at most equal to the total number of occurrences, we see that

$$0 \le p(E) \le 1.$$

If $p(E) = 0$ then E is impossible; if $p(E) = 1$ then E is certain.

The techniques of combinatorial probability centre around ways of combining events. Suppose we have two distinct events E and F. What is the probability that *either* E or F occurs?

Take the case of a die. E is the event '6 is thrown' and F the event '5 is thrown'. E or F is '5 or 6 is thrown' which obviously occurs 2 times out of 6. So

$$p(E \text{ or } F) = 1/3.$$

In general, let $N(E)$ and $N(F)$ be the number of ways in which E and F can occur, and T the total number of occurrences. Then

$$p(E \text{ or } F) = N(E \text{ or } F)/T.$$

What is $N(E \text{ or } F)$? Suppose the events E and F do not 'overlap'. (I'll return to this point.) Then

$$N(E \text{ or } F) = N(E) + N(F)$$

so that

$$
\begin{aligned}
p(E \text{ or } F) &= (N(E) + N(F))/T \\
&= (N(E))/T + (N(F))/T \\
&= p(E) + p(F). \quad (1)
\end{aligned}
$$

If, however, E and F do overlap then $N(E) + N(F)$ counts everything in the overlap *twice* whereas $N(E \text{ or } F)$ only counts it *once*.

Suppose, for instance, that

$$E = \text{'A prime number is thrown'}$$
$$F = \text{'An odd number is thrown'}.$$

Then E occurs in three ways: 2, 3, 5. (Note: 1 is not prime.) And F occurs in three ways: 1, 3, 5. But E or F occurs in *four* ways: 1, 2, 3, 5. So

$$p(E) = 1/2 \qquad p(F) = 1/2 \qquad p(E \text{ or } F) = 2/3.$$

What happens in general is that

$$N(E \text{ or } F) = N(E) + N(F) - N(E \text{ and } F) \quad (2)$$

because subtracting $N(E \text{ and } F)$ puts right the double count in the overlap. In the above example, E and F occurs in two ways: 3, 5. So the equation gives

$$4 = 3 + 3 - 2$$

which is correct.

Dividing (2) by T we get

$$p(E \text{ or } F) = p(E) + p(F) - p(E \text{ and } F). \quad (3)$$

Enter Set Theory

We can express these ideas much better in terms of sets. The possible *outcomes* when throwing a die form a set

$$X = \{1, 2, 3, 4, 5, 6\}.$$

The events E and F are represented by *subsets* of X

$$E = \{2, 3, 5\}$$
$$F = \{1, 3, 5\}$$

as in Figure 172.

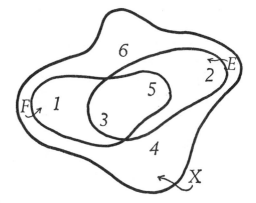

Figure 172

The event '*E* or *F*' is the set {1, 2, 3, 5} which is the *union* $E \cup F$. The event '*E* and *F*' is the set {3, 5} which is the intersection $E \cap F$. The probability p is a function defined on the set $\mathscr{E}$ of all subsets of *X* with target **R**. In general we can say a little more about p: it has target [0, 1], where this denotes the set of real numbers between 0 and 1.

Abstracting from this we obtain the idea of a *finite probability space*. This comprises

    (i) a finite set *X*,
    (ii) the set $\mathscr{E}$ of all subsets of *X*
    (iii) a function $p: \mathscr{E} \to [0, 1]$ with the property that
$$p(E \cup F) = p(E) + p(F) - p(E \cap F)$$
for all $E, F \in \mathscr{E}$.

Axiomatic probability theory works entirely in terms of probability spaces. However, if one wishes to consider infinite probability spaces, the definition has to be made more subtly. In many applications it is necessary to have infinite sets *X*: for example the height of a man can be any real number (within certain limits) so there are infinitely many possibilities.

Independence

Another basic operation in probability theory deals with two trials in succession: what is the probability of event *E* occurring on the first trial, event F on the second? For example, we throw a die twice: what are the chances that we throw first a 5, then a 2?

Of the 36 possible combinations, only 1 is favourable: 5 followed by 2. So the probability is 1/36.

If E and F were the events considered in the previous section, then E can occur first in 3 ways, F second in 3 ways. We can pair any occurrence of E with any of F, giving $3 \times 3 = 9$ favourable outcomes. So the probability of E followed by F is 9/36 = 1/4.

In general we must suppose that there are T_1 possible outcomes of the first trial, of which $N(E)$ are occurrences of E; and T_2 in the second, $N(F)$ being occurrences of F. Then in the two trials together the total number of outcomes is $T_1 \times T_2$, because any of the T_1 possibilities for the first can be followed by any of the T_2 possibilities for the second. In the same way the number of ways in which E can occur first, followed by F, is $N(E) \times N(F)$. So

$$p(E \text{ followed by } F) = \frac{N(E) \times N(F)}{T_1 T_2}$$

$$= \frac{N(E)}{T_1} \times \frac{N(F)}{T_2}$$

$$= p(E) \times p(F). \qquad (4)$$

In this calculation we must assume that E and F are *independent*: that the outcome of the first trial does not alter the probabilities in the second one.

This would not be the case if, say, the second event F was 'the total thrown is 4'. For if the first throw is 4 or more, the chance of success on the second is 0; if the first throw is 1, 2, or 3 the chance of success on the second is 1/6.

The notion of independence can be formulated in terms of probability spaces. In applications, one takes as hypothesis the independence of the real-world events to be considered, applies the theory, and tests the result by experiment.

Paradoxical Dice

Often our intuition about probabilities is wrong. Consider four dice A, B, C, D marked

A: 0 0 4 4 4 4
B: 3 3 3 3 3 3
C: 2 2 2 2 7 7
D: 1 1 1 5 5 5.

(The precise arrangement of faces does not matter.)

What is the probability that in a single throw die A will have a higher number showing than die B?

B always throws a 3. If A throws 4, which happens 4 times out of 6, he wins. If he throws 0, which happens 2 times out of 6, he loses. Therefore

A *beats* B *with probability* 2/3.

If B is thrown in competition with C it will win when C shows 2, lose when C shows 7. So

B *beats* C *with probability* 2/3.

If C plays against D matters are more complicated. With probability 1/2, D shows 1, and then C always wins; with probability 1/2 D shows 5 and C wins by showing 7 with probability 1/3. The probability that C will win is therefore

$$\tfrac{1}{2}.1 + \tfrac{1}{2}.\tfrac{1}{3} = \tfrac{1}{2} + \tfrac{1}{6} = \tfrac{2}{3}.$$

Thus

C *beats* D *with probability* 2/3.

Finally, look at D versus A. If D shows 5, with probability 1/2, then D always wins. If D shows 1, with probability 1/2, then D wins if A shows 0, which has probability 1/3. The probability that D will win is

$$\tfrac{1}{2}.1 + \tfrac{1}{2}.\tfrac{1}{3} = \tfrac{2}{3}.$$

Thus

D *beats* A *with probability* 2/3.

Now a die which wins more often than not is clearly 'better' than one which loses more often than it wins. In these terms,

A is better than B
B is better than C
C is better than D
and D is better than A.

There is nothing wrong with these calculations. If you play the game in practice, and let your opponent choose his die, then you can always choose another that gives you odds of 2:1 for a win.

We expect that A better than B better than C better than D should mean A better than D. We're wrong. In the present context the meaning of 'better than' depends on the choice of dice: we are really playing four *different* games. It is as if we had four people playing games: Alfred beats Bertram at tennis, Bertram beats Charlotte at chess, Charlotte beats Dierdre at badminton – and Dierdre beats Alfred at shove-halfpenny.

Those economists who believe that commodities can be ordered by majority preference might take note of this phenomenon.

Binomial Bias

Imagine a *biased* coin. Instead of coming down heads and tails with equal frequency, it has a preference for one particular side.

Such a coin provides a model for many probabilistic processes. If we are throwing a die and are interested only in whether a 6 turns up, we are effectively dealing with a biased coin such that $p(\text{head}) = 1/6, p(\text{tail}) = 5/6$. If we are looking at the sex of new-born babies, we have $p(\text{boy}) = 0.52, p(\text{girl}) = 0.48$.

In general we let

$$p = p(\text{head})$$
$$q = p(\text{tail})$$

and of course $p+q = 1$, because from (1) above

$$p(\text{head})+p(\text{tail}) = p(\text{head or tail}) = 1.$$

Using the theory of independent events we easily find the following list of probabilities for sequences of heads and tails:

H	p	HH	p^2	HHH	p^3
T	q	HT	pq	HHT	p^2q
		TH	pq	HTH	p^2q
		TT	q^2	HTT	pq^2
				THH	p^2q
				THT	pq^2
				TTH	pq^2
				TTT	q^3.

What is the probability that we throw a given number (0, 1, 2, or 3) of heads? We have to group together sequences with the same number of heads. Thus for 2 heads in 3 throws we get

HHT, HTH, THH, each with probability p^2q, which gives a total probability of $3p^2q$. Similar calculations give another table:

number of heads

		0	1	2	3
number	1	q	p		
of	2	q^2	$2pq$	p^2	
throws	3	q^3	$3pq^2$	$3p^2q$	p^3

The rows of this table should look familiar: compare the expansions

$$(q+p)^1 = q+p$$
$$(q+p)^2 = q^2+2pq+p^2$$
$$(q+p)^3 = q^3+3pq^2+3p^2q+p^3.$$

The terms on the right are exactly the entries in the table. The next row ought to come from

$$(q+p)^4 = q^4+4pq^3+6p^2q^2+4p^3q+p^4$$

and it is good practice to check that it does. In general, the entries in the nth row will be the terms of the expansion of

$$(q+p)^n.$$

This is not a coincidence, and it is not difficult to explain it. To expand, say, $(q+p)^5$ we must work out

$$(q+p)(q+p)(q+p)(q+p)(q+p).$$

The terms with exactly 3 qs come from products like this:

$$
\begin{array}{ccccc}
q & q & q & p & p \\
q & q & p & q & p \\
q & q & p & p & q \\
q & p & q & q & p \\
q & p & q & p & q \\
q & p & p & q & q \\
p & q & q & q & p \\
p & q & q & p & q \\
p & q & p & q & q \\
p & p & q & q & q.
\end{array}
$$

These correspond exactly to the 10 possible sequences of 3 tails and 2 heads:

$$
\begin{array}{ccccc}
T & T & T & H & H \\
T & T & H & T & H \\
T & T & H & H & T \\
T & H & T & T & H \\
T & H & T & H & T \\
T & H & H & T & T \\
H & T & T & T & H \\
H & T & T & H & T \\
H & T & H & T & T \\
H & H & T & T & T.
\end{array}
$$

Obviously the same holds in general. If we write $\binom{n}{r}$ for the number of sequences of n Hs and Ts containing exactly r Hs and $(n-r)$ Ts, then the probability of getting exactly r heads in n throws is

$$
\binom{n}{r} p^r q^{n-r}.
$$

It isn't too hard to work out what $\binom{n}{r}$ is. If we choose the r positions for Hs then everything is determined; so $\binom{n}{r}$ is just the number of ways of choosing r things from n. This, it can be shown, is given by

$$
\binom{n}{r} = \frac{n(n-1)(n-2)\dots(n-r+1)}{r(r-1)(r-2)\dots 1}.
$$

Thus for sequences of 2 heads and 3 tails we want

$$
\binom{5}{2} = \frac{5\cdot 4}{2\cdot 1} = 10
$$

which is correct.

The general expansion is

$$
(q+p)^n = q^n + npq^{n-1} + \dots + \binom{n}{r} p^r q^{n-r} + \dots + p^n.
$$

This is the *Binomial Theorem*, usually credited to Isaac Newton. It may or may not be coincidence that at one period Newton was Master of the Royal Mint.

The average number of heads obtainable in n throws can be calculated from this formula, and it turns out to be np. Thus the frequency with which heads occur is np/n, which is p. So we have come full circle to the idea of a probability as an 'average frequency of occurrence'. This theorem, which in a stronger form is called the *Law of Large Numbers*, shows how our mathematical model connects up with observations in the real world.

Random Walks

In the final section of this chapter I want to discuss another type of problem arising in probability theory. It has applications to questions about electrons bouncing around inside crystals, and particles floating in a liquid.

Imagine a particle starting at position $x = 0$ on the x-axis, at time $t = 0$. In time $t = 1$ it moves to the point $x = -1$ with probability 1/2, or to the point $x = +1$ with probability 1/2. If it is in position x at time t, then at time $t+1$ it moves either to $x-1$ or $x+1$, each with probability 1/2. What can we say about the particle's subsequent motion?

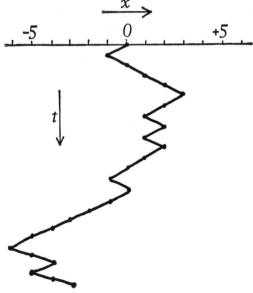

Figure 173

For example, it may move left and right according to the sequence

$$LRRRRLLRLRLLLRLLLLLLLRRLRR$$

in which case its motion is as in Figure 173, with the path stretched a little in the t direction for clarity. This is a fairly typical path: if you want, you can use a coin to decide between L and R and construct other paths for yourself.

Instead of a line we could consider a plane: now the particle moves either 1 unit up, down, left, or right; each with probability 1/4. Or a 3-dimensional walk, with 6 possible directions each having probability 1/6.

Especially interesting is the question: given any other point X, what is the probability that eventually (we don't mind how long it takes) the particle reaches X?

One would expect this to decrease as X gets further away from the origin. In fact it does nothing of the kind: the probability remains the same for *all* X. In a random walk every point is as good as every other point, in the long run.

For a random walk in 1 or 2 dimensions, this probability is 1. It is almost certain that the particle will reach any given point X. (I say 'almost' because it may not do so: it could rush off to the right and never be seen again. But this has probability 0. With infinite processes, it is not quite true[2] that probability 1 means 'certain' and 0 'impossible'.)

But in 3 dimensions, the probability is only 0·24.

If you were lost in 1- or 2- dimensional space, and wandered about at random, then with probability 1 you would eventually find your way home. In 3 dimensions your chances of getting home are less than 1 in 4.

However, in all cases, the average time it would take you to arrive home is *infinite*. More precisely, pick any time t_0 – it might be 5 seconds or 3000 years. Then if you keep wandering, on most occasions you will be away from home for a time larger than t_0.

Chapter 18   Computers and Their Uses

'A hydrodynamicist was reading a research paper translated from the Russian, and was puzzled by references to a "water sheep". It transpired that the paper had been translated using a computer: the phrase in question should have read "hydraulic ram" ' –
Cautionary tale

Strictly speaking, computing is not part of mathematics, but a discipline in its own right. The computer is not a *concept* of modern mathematics: it is a *product* of modern technology. Nevertheless most modern mathematics courses in schools include a certain amount of computing, and quite rightly so; for the computer is a powerful tool of great practical importance in the applications of mathematics to the modern world.

On the whole computers play no role in theoretical mathematics. In order to be able to put a problem on to a computer one must *in principle* know exactly how to perform the steps necessary to solve it. From the theoretical viewpoint this means that the problem is as good as solved already, especially if the main concern is *method*. But to get results (which in practical applications are of course the main desideratum) a method which works in principle is not enough: it must also work in practice. The importance of the computer lies in its ability to bridge the gap between principle and practice.

The computer is also of interest to the mathematician because of the mathematical ideas which underly its construction.

I want, in this chapter, to give some small idea of both the mathematical and practical considerations behind the design and use of computers. For the technical details the reader must consult a specialist book.[1]

Binary Notation

Basically the computer is a calculating machine. That is to say it is given data, in the form (usually) of numbers, told how to manipulate these data, and it then prints out the results. Most present-day computers are electronic digital computers: they use electronic circuitry to store and manipulate numbers in digital form. However there are other possibilities: among them optical computers (which use beams of light) and fluidic computers (which use streams of liquid or gas). And in a moment I intend to illustrate some of the ideas behind the design of a computer in terms of a machine which uses ball-bearings.

A computer cannot operate directly with numbers, since numbers as such do not exist in the real world. It is necessary to represent numbers in the machine in some physical fashion. In an *analogue* computer a number x is represented by x units of current: but the problem of maintaining accuracy, the inflexibility of such a machine, and its slowness of operation, render it fit only for a limited range of tasks. Something more subtle is needed.

The simplest devices capable of representing numbers are those which can exist in either of two stable states. A switch may be on or off. A current may either be flowing or not flowing. A magnet may be magnetized north–south or south–north. The possibility of using such devices to store and manipulate numbers is opened up by the existence of the *binary notation.*

In everyday arithmetic we represent numbers by *decimal* notation. Thus

$$365 = (3 \times 10^2) + (6 \times 10) + (5 \times 1)$$
$$1066 = (1 \times 10^3) + (0 \times 10^2) + (6 \times 10) + (6 \times 1).$$

The occurrence of the various powers of 10 here is one of choice rather than necessity. There is no special reason why we have to use 10: we could use, say, 6 instead. The sequence of numbers would then be

$$1 = (1 \times 1)$$
$$2 = (2 \times 1)$$
$$3 = (3 \times 1)$$
$$4 = (4 \times 1)$$
$$5 = (5 \times 1)$$
$$10 = (1 \times 6) + (0 \times 1)$$
$$11 = (1 \times 6) + (1 \times 1)$$
$$12 = (1 \times 6) + (2 \times 1)$$

...

$$55 = (5 \times 6) + (5 \times 1)$$
$$100 = (1 \times 6^2) + (0 \times 6) + (0 \times 1)$$

. . .

Such a system might be evolved by creatures having only 6 fingers.

The simplest notational system of this kind is the *binary* system, which uses powers of 2. This is the system we would have evolved had we counted on our 2 hands instead of our 10 fingers. The only digits needed are 0 and 1, and the sequence of numbers runs

$$1 = \qquad\qquad (1 \times 1) \quad [= 1]$$
$$10 = \qquad\qquad (1 \times 2) + (0 \times 1) \quad [= 2]$$
$$11 = \qquad\qquad (1 \times 2) + (1 \times 1) \quad [= 3]$$
$$100 = (1 \times 2^2) + (0 \times 2) + (0 \times 1) \quad [= 4]$$
$$101 = (1 \times 2^2) + (0 \times 2) + (1 \times 1) \quad [= 5]$$
$$110 = (1 \times 2^2) + (1 \times 2) + (0 \times 1) \quad [= 6]$$
$$111 = (1 \times 2^2) + (1 \times 2) + (1 \times 1) \quad [= 7]$$
$$1000 = (1 \times 2^3) + (0 \times 2^2) + (0 \times 2) + (0 \times 1) \quad [= 8]$$
$$1001 = (1 \times 2^3) + (0 \times 2^2) + (0 \times 2) + (1 \times 1) \quad [= 9]$$

. . .

(where the figures in square brackets give the number in ordinary decimal notation).

The ordinary methods of adding, subtracting, multiplying, and dividing apply in this notation; except that anything larger than 1 is 'carried'. The necessary information for addition is given by

$$0 + 0 = 0$$
$$1 + 0 = 1 \qquad\qquad (+)$$
$$0 + 1 = 1$$
$$1 + 1 = 0 \text{ carry } 1.$$

The multiplication table is even easier:

$$0 \times 0 = 0$$
$$0 \times 1 = 0 \qquad\qquad (\times)$$
$$1 \times 0 = 0$$
$$1 \times 1 = 1.$$

Children brought up with this notation would have little trouble learning their tables!

All of arithmetic can be carried out on the basis of the tables $(+)$ and $(\times)$. For example, to multiply 11 011 by 1 010 by standard long multiplication we write

$$
\begin{array}{r}
11\ 011 \\
1\ 010 \\
\hline
11\ 011\ 000 \\
110\ 110 \\
\hline
100\ 001\ 110 \\
\hline
{\scriptstyle 1\ 11\ 1}
\end{array}
$$

(where the small ₁s are carry digits).

As a check you should note that in decimal notation 11 011 = $16+8+2+1 = 27$, $1\ 010 = 8+2 = 10$, and $100\ 001\ 110 = 256+8+4+2 = 270$.

The point is that binary and decimal arithmetic differ only in choice of *notation*: they are both about the same kinds of number.

A Ball-Bearing Computer

I want to illustrate how the tables $(+)$ and $(\times)$ can be realized by a machine. In order to take our minds off the wizardry of electronics I shall show how to use ball-bearings to make an adding machine. The general principles are the same for an electronic computer, except that it uses pulses of electricity instead of ball-bearings.

First we must design a component which behaves according to the table $(+)$. It should have two stable states (which for convenience we denote by 0 and 1) and should react to an 'input', also taking values 0, 1, as follows:

input	initial state	final state	output
0	0	0	0
0	1	1	0
1	0	1	0
1	1	0	1

(The initial state of course represents one of the digits being added; the input is the other digit; the final state is the sum digit; the output the carry digit.)

If we use 1 ball-bearing to represent an input of 1, and 0 ball-bearings to represent an input of 0. we can make an 'adder' as in Figure 174 using a T-shaped component pivoted at the junction of upright and crossbar. Gravity will provide the motive force.

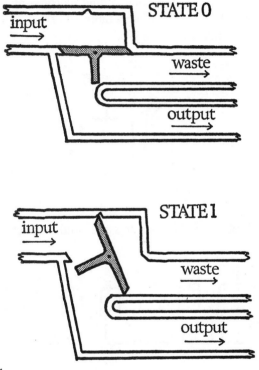

Figure 174

It will be seen that

(i) If the device is in state 0 and input 0 (i.e. no ball-bearing) is applied, it stays in state 0.

(ii) Similarly if it is in state 1 and 0 is input it stays in state 1.

(iii) If it is in state 0 and 1 is input by rolling a ball-bearing through, then the T tips up to state 1. The ball-bearing emerges through the 'waste' channel: the output is 0.

(iv) If it is in state 1 and 1 is input the T tips back to the 0 state. But the ball-bearing emerges through the output channel, giving output 1.

Thus our device does exactly what is required.

Now we can build a full-scale adding machine by combining together several of these 'adding units'. Let me represent the device by the symbol of Figure 175.

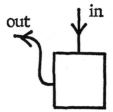

Figure 175

The 'circuit' of Figure 176 then acts as an adding machine.

Figure 176

To perform (say) the addition of 11 011 000 and 110 110 (which we did in the multiplication above) we set up the first number in the machine and input the second by means of ball-bearings, as shown schematically in Figure 177.

If we imagine the ball-bearings added one by one, from right to left, we can follow the steps of the calculation. There is no ball-

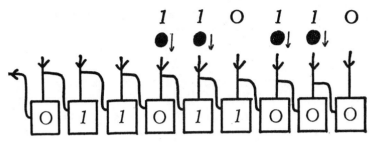

Figure 177

bearing in the first slot. The one in the second slot changes 0 to 1 and is output as waste, leaving the machine in states

0 1 1 0 1 1 0 1 0.

The same happens with the third slot:

0 1 1 0 1 1 1 1 0.

The fourth slot has no input. The fifth has one ball-bearing, which passes through, changes the 1 to a 0, and is output as a carry into column 6:

0 1 1 0 0 1 1 1 0.
$\quad\quad$ 1 ←

Then it passes through again, changing 0 to 1 and emerging as waste:

0 1 1 1 0 1 1 1 0.

Finally the ball-bearing in column 6 drops through, making a series of changes as follows:

0 1 1 0 0 1 1 1 0
$\quad\quad$ 1 ←

0 1 0 0 0 1 1 1 0
$\quad\quad$ 1 ←

0 0 0 0 0 1 1 1 0
$\quad$ 1 ←

1 0 0 0 0 1 1 1 0.

This is the correct answer.

The reader should check that the sequence of operations above corresponds exactly to what happens in the addition sum, and also try a few other examples. An interesting practical problem is to find the correct spatial arrangement of the components to make the machine work using only gravity as a motive force. This can be done.

Of course in an electronic computer one uses pulses of electricity instead of ball-bearings and electronic components instead of T-shaped barriers. But the underlying idea is similar.

Multiplication can be carried out too (as a series of additions, for example). By the use of a small number of basic circuits, repeated many times, it is possible to build a versatile and accurate calculating machine. Since electronic circuitry reacts very quickly, it will also be a fast machine.

The Structure of Computers

The ideas we have just touched upon allow the construction of an *arithmetic unit*. However, there is more to a computer than this, for an arithmetic unit alone has no flexibility. The basic structure of a computer is shown below.

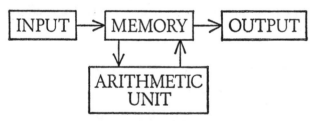

The memory of the computer has two functions. First, it stores the numbers which are input, which crop up in the calculation, or are about to be output. Secondly, it stores the *program* (the American spelling is standard in this field) which tells the machine what steps to make in the calculations. The computer, so to speak, looks up an instruction in the program, performs it, remembers the answer; then it looks up the next instruction, and so on.

In the first instance these instructions are in *machine language* – a special 'code' which the machine 'understands'. These are of a very specialized and precise nature. 'Remove the contents of position 17 in the memory and put them in the arithmetic unit.' 'Add the two digits in the arithmetic unit.' And suchlike. Even to multiply two numbers together in machine language requires many instructions.

For this reason other programming languages have been devised which are closer to ordinary language: an instruction such

$$C = A+B$$

would tell the machine to add the numbers stored under the names of A and B and store the result under the name of C. The machine has to be provided with a *compiler* program, written in machine language, which turns each such higher-language instruction into a series of machine-language instructions.

There are many kinds of higher language, rejoicing under names such as Algol, Fortran, Cobol (a commercial language). Machines are delivered from the manufacturers together with the necessary compiler programs.

The use of a program is what gives the computer its great flexibility. It will carry out any sequence of instructions which the programmer gives it. In consequence a great many different tasks can be carried out by the same machine. The programmer has to learn one or more of the standard languages, which is not too hard. Much harder is the art of programming: how to use the languages efficiently and effectively.

Writing a Program

Assuming you have learnt a suitable language, and have a problem you wish to solve using a computer, how do you go about writing a program?

The first step is to break the problem down into small pieces, each of which the computer can perform, and then to write a program organizing the pieces.

For example, you wish to solve quadratic equations. You know that the answer to

$$ax^2+bx+c = 0$$

is given by

$$x = \frac{-b + \sqrt{(b^2 - 4ac)}}{2a}.$$

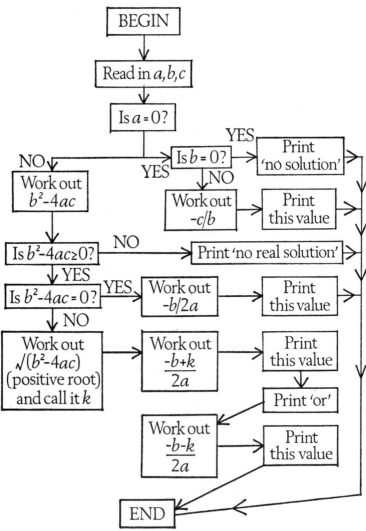

Figure 178

It is no good just putting this instruction into the computer. It might happen that (b^2-4ac) is negative: the computer will not realize this and will try to find the square root, with nonsense as a result. Or a might be 0, in which case the division will not make sense.

Assume that the computer can do arithmetic, including taking square roots, and can recognize whether a number is positive or negative. You might break the calculation down as in Figure 178. Such a diagram is called a *flow-chart*.

You will see that this chart takes account of several possibilities: if $a = 0$ we may have a linear equation; there may be no real roots, or 1 real root, or 2 real roots.

The next step is to convert the process on the chart into a program. There is a difficulty here, in that the program is a sequence of instructions in order, whereas the chart has bifurcations and alternatives. To overcome this, different parts of the program are given reference letters (A, B, C, D, E in the following example) and the machine is instructed to jump from one part to another depending on whether the answer to certain questions is Yes or No.

Here is a possible program for the above calculation, written in a hypothetical language based on Algol. It is largely self-explanatory, if considered in conjunction with the flow-chart.

PROGRAM TO SOLVE QUADRATIC EQUATIONS

```
A: begin real a, b, c, k, u, v, w, x, y
        read a, b, c
        if a = 0 then go to B
        y = b²-4ac
        if y ≥ 0 then go to C
        print NO REAL SOLUTION
        end

B:      if b = 0 then go to E
        x = -c/b
        print x
        end
```

C: if $y = 0$ then go to **D**
$k = \sqrt{y}$
$u = (-b+k)/2a$
$v = (-b-k)/2a$
print u
print **OR**
print v
end

D: $w = -b/2a$
print w
end

E: print **NO SOLUTION**
end

The instruction 'real $a, b, c, k, u, v, w, x, y$' is known as a *declaration*, and tells the computer what symbols are being used to represent numbers and what kind of number (real). 'Read' instructs it to read in the values of a, b, c from some prepared data tape. An instruction of the form 'if P then go to X' is obeyed as follows: if P is true the machine jumps to the part of the program labelled X. Otherwise it goes on to the next line. The other instructions are self-explanatory: the computer obeys them line by line in order except when instructed to jump.

This is a fairly typical simple program which I hope may help to explain how a program is written. You should work through it with a few specific examples of values of a, b, c to see how it does the job.

For further details of programming in proper languages, you should consult a suitable manual.[2]

The Uses of Computers

Computers can be used, quite simply, whenever one has a lot of calculation to do of a kind which can be specified precisely. Whether or not they *should* be used is often a question of economics: they are expensive; is the result worth the expense?

Computers are used in business and government, mostly to keep accounts and to file information. Of these uses I shall say no more except to note in passing that the much-pleaded excuse 'computer error' is in reality 'programmer error'.

The research worker can make good use of computers to process his experimental data, plot graphs, calculate tables of results, apply statistical techniques. He can solve otherwise intractable equations by numerical methods. There is a danger that one can become so impressed by the sheer bulk of information obtained by computing that one fails to realize that much of it may be valueless: no amount of computing can produce useful results from a badly designed questionnaire or experiment. But the power which the computer lends is enormous. Its use has given us insights into the structure of proteins, the genetic code, the fundamental particles of physics, the structure of stars. It has helped to put a man on the moon.

Even in pure mathematics the computer has scored some notable triumphs, especially in the study of finite groups. However, very few problems are suitable for computation; and even some of those that are would take too long to perform, even for today's very fast machines (or tomorrow's, for that matter).

The uses of computers are not confined to numerical problems. Computers have been programmed to play draughts (well) and chess (badly), to translate from one language to another (execrably), to compose music (of sorts) and poetry. Some of the recent advances in producing 'intelligent' machines are quite remarkable.

This brings me naturally to the oft-asked question 'Can computers think?' As Joad would have said, it all depends what you mean by 'think'. As yet, the computer can perform some of the functions of the human brain faster and more accurately; others it cannot perform at all. But if we ask, 'Is there something special about the way in which human beings think which *in principle* can never be performed by some kind of machine?' then my personal opinion is that the answer is 'No'. Certainly we cannot duplicate the functions of the brain at the present time; and it is fairly certain that the resemblance between the brain and existing computers is about as close as that between a cow and a milk-lorry. Our technology may well never get anywhere near making a truly 'intelligent' machine: the human brain may well be too stupid. But I don't think there is any obstacle to the production of a machine which performs the functions of the human brain; not any logical obstacle such as prevents $\sqrt{2}$ from being rational or a man from lifting himself by his bootstraps; for the following reason: the human body is *visibly* a machine, in the sense that it composed out of matter and the components obey the same laws as other matter. It is a very complicated and wonderful machine which we don't understand. If there were in principle an obstacle to the construction of machines which behaved like people, then there would be no people.

This is not to reduce humanity to the level of a can-opener. Many people insist that the complexities of human behaviour, the emotional, creative, and spiritual attributes, must be consequences of something 'greater' than physical laws. This is a wonderful concept. How much more wonderful it would be, however, if these very attributes *were* consequences of physical laws. Far from demeaning humanity, this would elevate physics!

Chapter 19   Applications of Modern
                 Mathematics

Although I have talked in terms of divisions of mathematics into various branches – algebra, topology, analysis, logic, geometry, number theory, probability – it must be realized that no exact boundaries can be drawn, and that the divisions themselves are somewhat arbitrary. When Descartes first discovered connections between geometry and algebra it was a surprise. When Galois applied the theory of groups to polynomial equations it was a surprise. When Hadamard and de la Vallée Poussin proved an important conjecture about prime numbers using analysis it was a surprise. Today mathematicians are no longer surprised by such occurrences. In fact they tend to go out looking for them. It is quite commonplace to start with a problem in analysis, turn it into topology, reduce it to algebra, and solve it by number theory.

It is this unity which makes it feasible to talk about the 'central body of mathematics' as I did in Chapter 1. The subject is so interconnected that a genuine advance in any part of this central body is of importance for the whole of mathematics. Mathematics is a harmonious whole: except that the harmony is incomplete, because there are always gaps in our knowledge and vaguely understood hints of new interrelations.

In this sense, an application of any mathematics from this central body is an application of the whole. If you insist that mathematics justify its existence by providing applications, then an application of one part will justify the whole. We do not cut off a violinist's feet just because he doesn't use them in playing the violin: in the same way we ought not to dismiss group theory just because it won't pay the rent.

Traditionally there are two kinds of mathematics: pure, and applied. The pure mathematician has his head in the abstract clouds and studies the subject for its own sake: he is not interested in applications, and if anything tends to denigrate them. The applied mathematician has his feet on the concrete and provides a useful service to society.

Like most traditions, there is a grain of truth in this one. Mathematics is such a large subject that workers are forced to specialize in some small part. If this part has no direct applications to the real world they are placed in the 'pure' category: if it does have applications they are placed in the 'applied' category. But a great deal of supposedly pure mathematics has important applications, and a great deal of supposedly applied mathematics has no *useful* applications at all. I am reminded of a man who developed a mathematical theory of the paintbrush. To set up equations he could solve he had to assume that the bristles of the brush were semi-infinite planes. So his theory gave no insight whatever into paintbrushes; and precious little into mathematics because he deliberately set up equations that could be solved by known methods.

I would prefer to say that there is (i) mathematics, (ii) applications of mathematics. The job of the mathematician is to provide powerful tools for solving mathematical problems: these may be suggested by possible applications, they may be part of a more abstract investigation into stumbling-blocks in mathematical technique, or important unsolved problems. As I said in Chapter 1, there is a large time-lag in applications of mathematics: the pure mathematics of one century may be the theoretical physics of the next. Of course applications are important; but it is no good taking too short-term a view of them.

I want to give three instances of important applications of modern mathematics. The first shows how linear algebra can be used to solve certain kinds of problems in economics, and is about as close as group theory gets to paying the rent. The second is a recent application of group theory to the study of fundamental particles in physics. The third is a brand new theory of discontinuous processes, which is based on brand new mathematics. This might eventually have important applications to biology and medicine; it has already been successfully used to study the propagation of nerve-impulses.

This last theory is so new that much of it is pure speculation and most of it has yet to be worked out. If I am to beat the time-lag effect I have to indulge in a little crystal-gazing. But if one recalls that the calculus is largely a study of continuous processes, and notes that the calculus has for two centuries been the basic tool of theoretical science; if one realizes that there is a tremendous need for insight into discontinuous processes in physics, chemistry, engineering, meteorology, biology, economics, sociology, politics, geophysics, aerodynamics, ..., then it should be at least plausible that a theory of discontinuous processes has considerable potential.

How to Maximize Profits

A certain factory manufactures two distinct products, 'gadgets' and 'doofers'. In each case the product is first turned on a lathe, and then has holes drilled in it. The times required for these operations, the total time available per week, and the profit per gadget or doofer are as tabulated:

machine	gadget	doofer	available time
lathe	3	5	15
drill	5	2	10
profit per unit	5	3	

How can the manufacturer make the most profit?

Let us suppose that he makes x gadgets and y doofers every week. Then time considerations yield conditions

$$3x+5y \leq 15 \tag{1}$$
$$5x+2y \leq 10 \tag{2}$$

while of course

$$x \geq 0 \tag{3}$$
$$y \geq 0. \tag{4}$$

His profit will be

$$5x+3y. \tag{5}$$

The problem is thus to maximize (5) subject to the system of inequalities (1)–(4).

We have no techniques for solving inequalities, so we draw a graph. The points (x, y) which satisfy conditions (1)–(4) are those lying in the shaded region of Figure 179. The last two conditions just tell us that x and y are positive; (1) says that (x, y) lies below the line $3x+5y = 15$; (2) that (x, y) lies to the left of the line $5x+2y = 10$.

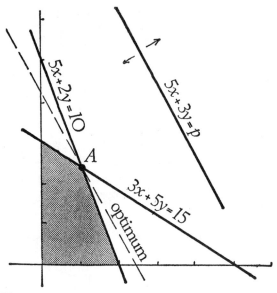

Figure 179

For a given profit p the line $5x+3y = p$ is as shown. As we change p the line moves, but its slope stays the same. The further the line moves to the right, the larger p becomes.

Our problem is to find the biggest value of p for which this line passes through the shaded region, because the shaded region represents the possible (x, y). This obviously occurs when the line passes through the point A at one corner of the shaded region.

To find A we solve

$$5x+2y = 10$$
$$3x+5y = 15$$

which give

$$x = 20/19 \qquad y = 45/19.$$

The profit per week is then $5x+3y$, which is

$$\frac{100+135}{19} = \frac{235}{19}.$$

The factory should produce 45 doofers and 20 gadgets every 19 weeks in order to make the maximum profit.

Similar considerations apply to any business, or national economy. However the number of products and machines will be very large. The general problem will be to maximize a certain linear combination of the unknowns, subject to a system of linear inequalities. These inequalities determine a certain region in a multidimensional space. It can be shown that

(i) this region is convex;
(ii) the maximum profit occurs, if it exists, at one of the corners of the region.

The proof involves essential use of linear algebra. So does the technique needed to find the corner at which the maximum occurs: and with a lot of unknowns a computer is required.

To study a national economy in complete detail would involve so many equations that even the fastest computer available could not hope to handle them. Some simplifying assumptions must then be made; with a consequent doubt as to the validity of the results.

This technique, which is known as *linear programming*, is a standard one of economics, and may be found in most textbooks on mathematical economics.[1]

The Eightfold Way

At one time atomic theory was fairly simple. All atoms were thought to be made up of three different kinds of fundamental particle: protons, neutrons, and electrons. More penetrating investigations revealed the existence of hosts of other fundamental particles: neutrinos, pions, muons, and the like. But there was no theory available to organize the various particles into a coherent structure.

In 1964 it was discovered that group theory could be used to provide such an organization. The following description of the idea is of necessity highly compressed, so do not expect to be able to follow more than a general outline.

The basic technique involves *representations* of groups. Given a group G we find its representations as follows: we look for a vector space V having some linear transformations which form a group G' isomorphic to G. This G' (or more strictly the isomorphism) is said to be a representation of G.

Consider the example where G is the group with two elements $\{I, r\}$ such that $r^2 = I$. If we take $V = \mathbf{R}^2$ we can consider a reflection T in some fixed line through the origin. If we let I be the identity map, then $\{I, T\}$ forms a group of linear transformations of V. Furthermore $T^2 = I$, so that $G' = \{I, T\}$ is isomorphic to G.

The dimension of the space V is, by abuse of language, referred to as the dimension of the representation.

In quantum mechanics a given object may exist in various different energy states. In a hydrogen atom, consisting of a proton and an electron, the electron may take energies belonging to a certain infinite, but precisely determined, set of values. The electron can change state by absorbing or emitting photons to keep the total energy constant.

There is a mathematical consequence of the laws of quantum mechanics: the possible states of a physical object correspond precisely to the *representations* of the *symmetry group* of the object.

For example, a single atom floating in empty space at a fixed point P has complete rotational symmetry: its symmetry group is the group O_3 of all rigid motions of 3-space that leave P fixed. This group already *is* a group of linear transformations of 3-space, because rigid motions are linear transformations, so it has a 3-dimensional representation (which the physicists call the *triplet* representation).

If now we turn on a magnetic field the symmetry is destroyed: the direction of the field defines a line in 3-space, and the symmetry group is now the group O_2 of rotations keeping this line fixed. It turns out that the triplet representation of O_3 splits up into 3 different 1-dimensional representations of O_2. In a spectroscope the single spectral line which occurs in the absence of a magnetic field splits into 3 closely spaced lines when the field is turned on. The energies can be calculated, and agree with experiment.

This use of group theory is quite common in quantum mechanics. It was used in 1938 to predict the existence, and various properties, of pions. Pions were detected experimentally in 1947 and the predicted properties held good.

Among the known fundamental particles there are some which are much more massive than the rest, known collectively as *baryons*. They include the neutron n^0, the proton n^+, and more esoteric particles denoted by Λ (lambda), Ξ (xi), Σ (sigma), and Δ (delta). Each particle has a certain mass, and an electric charge which always comes in integer multiples of the basic unit of charge, which is $+1$ for the proton and 0 for the neutron. (It is -1 for the electron but this is not a baryon.)

There are other physical quantities associated with fundamental particles, less intuitive than mass or charge. Among them are *spin, isotopic spin, hypercharge,* and *strangeness*.

The commonest baryons are eight in number, and comprise a Ξ doublet, a Σ triplet, a Λ singlet and an n doublet. Their mass, charge, isotopic spin (I) and hypercharge (Y) are as shown in Figure 180.

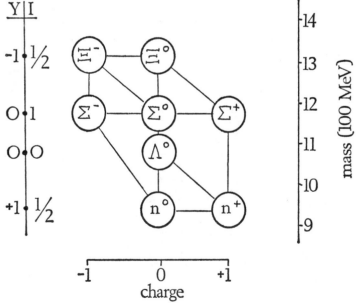

Figure 180

These can be organized in terms of the representations of a certain group called SU_3. The most natural representation of SU_3 has dimension 8. If in fact the SU_3 symmetry in nature is imperfect then the symmetry group reduces to a subgroup U_2. The original 8-dimensional representation splits into 4 parts, having dimensions 3, 2, 2, 1. These correspond exactly to the Σ triplet, the Ξ and n doublets, and the Λ singlet.

Further, the observed values of Y, I, mass, and charge agree with those predicted by the SU_3 theory. It is as if all the baryons are really different states of *one* fundamental particle, which has been perturbed into the eight different kinds by asymmetries in nature.

The theory is known as the 'eightfold way'.

A crucial test was made. The next representation of SU_3 has dimension 10. When restricted to U_2 it breaks up into 4 parts of dimensions 4, 3, 2, 1. Nine known particles fitted: a Δ quadruplet, a Σ triplet, and a Ξ doublet (Figure 181). (The masses of the Σs and Ξs differ from those in Figure 180 because we are looking at different states.)

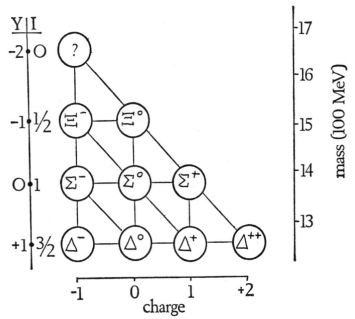

Figure 181

The question mark indicates a missing particle. The theory makes several predictions about this particle: it must have charge −1, hypercharge −2, isotopic spin 0, and mass about 1 700 MeV. As it happened, these were a highly unexpected combination.

In February 1964, in a specially designed experiment, the particle was found. It was dubbed the *omega-minus* (Ω^-).

A theory based on abstract groups had correctly predicted the existence of a hitherto unknown fundamental particle.[2]

Catastrophe Theory

It is not always true that continuous changes produce continuous effects. To switch on a light you move the switch in a smooth continuous path from the 'off' to the 'on' position; but there is a point in between at which the light *suddenly* switches from 'off' to 'on'. A continuous movement near the edge of a cliff can produce a discontinuous result if we fall off the edge.

Most of mathematics, and virtually all of physics, has until recently concentrated on continuous changes. However René Thom, one of the world's outstanding mathematicians, has discovered a profound theory of discontinuous changes, which he calls *catastrophes*.[3]

The potential applications of such a theory are widespread and important. Perhaps the most important is in the field of biology. As an embryo develops it passes through many discontinuous changes, as cells divide, limbs begin to form, nerves and bones and muscles develop. An insight into these processes could lead to enormous strides in biological understanding. Perhaps one day we might be able to use this understanding in medicine, particularly as regards deformities in children.

Such applications, if they are possible, are many decades or centuries away. But Thom's theory is the only one which gives any kind of insight into discontinuous processes. As such, it is worth developing.

The subsequent discussion will be much facilitated if the reader constructs, or at the very least imagines, a *Zeeman Catastrophe*

Machine,[4] as shown in Figure 182. It consists of a pivoted disc, to the edge of which are attached two equal lengths of elastic. One is fixed at a point *F*, the other is free to move.

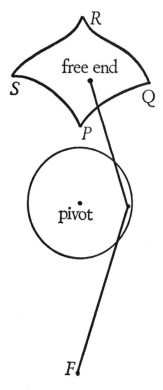

Figure 182

(Use a cardboard disc, and pin it to a wooden board with a drawing pin. With a 5-cm diameter disc the point *F* should be about 12 cm below the centre of the disc; the elastic should be about 8 cm long and not too strong. A washer helps. Preferably the elastic should be attached to the edge of the disc by some device which can rotate freely.)

Experiment will reveal the existence of a diamond-shaped region, roughly like *PQRS*, with the following property: outside the region, the disc can come to rest in only one position. Inside the region, there are two possible rest positions.

Furthermore, the disc can be made to jump suddenly from one rest position to a completely different one by a continuous variation of the free end of elastic. For a movement as in Figure 183 the disc jumps when the path moves out of (but *not* when it moves into) the diamond-shaped region *PQRS*.

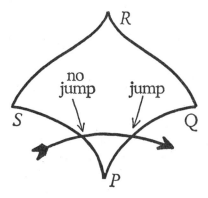

Figure 183

To see why this happens we analyse the energy in the elastic. One can imagine the disc artificially forced to some non-equilibrium position. When we let go, it twangs back to some position of equilibrium. It does this in order to minimize the energy stored in the elastic (more strictly to make the energy stationary, in a sense which will be explained below).

Outside the region *PQRS* the energy curve looks like Figure 184, where θ denotes the angle at which the disc is positioned. There is a single minimum, and thus a single equilibrium position.

Inside *PQRS* the energy curve looks like Figure 185. This time there are two minima, corresponding to different angles θ: this gives two equilibrium positions.

In between the minima is a maximum: this does in fact correspond to an equilibrium position, but an *unstable* one. The slight perturbation will cause the energy to 'roll down the slope' to a minimum. It is theoretically possible to balance a pin on its point: but such a position is one of unstable equilibrium.

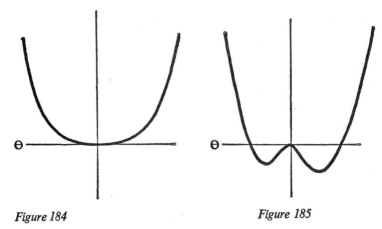

Figure 184   Figure 185

When we follow the path shown in Figure 183 the energy curve follows the sequence of changes of Figure 186.

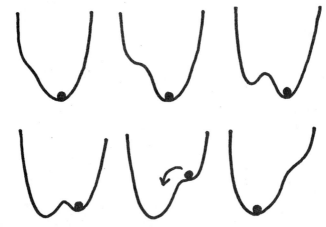

Figure 186

The disc starts off at one minimum, and because of continuity conditions it stays in that minimum, *all the while the minimum remains in existence*. When the minimum disappears, it cannot remain in it: so it moves to the only one left. The disc is trying to behave continuously, but it is forced to jump by circumstances beyond its control.

This behaviour can be seen more vividly if we draw a 3-dimensional graph showing, for each position of the free end, the possible equilibria. The result, it can be proved, is Figure 187.

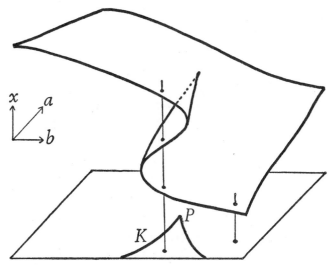

Figure 187

In this figure, the point P is shown, together with part of the diamond-shaped region. The region itself has been turned round for clarity. Over points inside the cusp-shaped line K there are three possible equilibria: one on the top layer of the fold, one in the middle, one below. The middle one is the unstable one. Over a point outside K, there is just one layer of surface.

As we follow a path across K the disc tries to stay in the equilibrium position corresponding to the top surface. However, when we pass out of K, it 'falls off the edge' and the disc jumps.

Quantitatively we proceed as follows. We take a certain system (a, b) of coordinates, with P at the origin. We also choose a variable x, related to the equilibrium angle of the disc. For small values of a, b, x the energy in the elastic, V, is

$$V = \tfrac{1}{4}x^4 + \tfrac{1}{2}ax^2 + bx.$$

To find the equilibria we want the *stationary values* of the energy. These are the points where the graph of the energy is horizontal:

maxima, minima, and 'points of inflection'. These possibilities are illustrated in Figure 188.

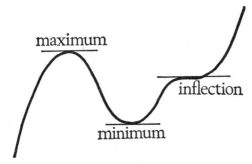

Figure 188

By using calculus it can be shown that the stationary values occur when $\dfrac{dV}{dx} = 0$, i.e.

$$x^3 + ax + b = 0.$$

If we plot the 'graph' of this equation by choosing particular values of a, b and calculating x we get Figure 187.

René Thom studied a general situation of which the above is a special case. He considered any dynamical system whose behaviour can be measured by variables

$$x, y, z, \ldots$$

and which is controlled by another set of variables

$$a, b, c, \ldots.$$

The variables x, y, z, , . . . are the coordinates of a *behaviour space*, and a, b, c, . . . are the coordinates of a *control space*. The behaviour of the system is governed by the *potential*, or energy, in the system: we take a completely general potential

$$V = V(x, y, z, \ldots, a, b, c, \ldots)$$

subject only to conditions which allow us to apply the operations of the calculus to V.

For fixed choices of a, b, c, . . . the system takes up equilibrium positions corresponding to stationary values of V.

In the Zeeman machine, we have one dimension of behaviour x, and two dimensions of control a, b: the potential V is as given.

Obviously we can obtain such a system for any choice of V, so there are infinitely many different systems of this kind. However, a lot of these become the same if we make 'changes of coordinates'. If in the Zeeman machine we change x to $2X$ then V becomes

$$4X^4 + 2aX + b$$

which is different from the old V. But if we knew all about the new system we would know all about the old one: the change is not a significant one. The easiest way to weed out insignificant variations is to look at the topological properties only.

All events in the physical world are controlled by 4 variables: 3 of space and 1 of time. So if we have physical applications in mind we may often restrict ourselves to 4 dimensions of control. (But it is not essential to do so.)

Thom then proved a wonderful theorem. With 4 dimensions of control there are precisely 7 topologically distinct kinds of discontinuity which typically occur in such a dynamical system. *Every energy-minimizing physical discontinuity is typically one of 7 basic types.*

Thom listed these 7 types, which he called the *elementary catastrophes*. They go like this:

name	potential V
fold	$\frac{1}{3}x^3 + ax$
cusp	$\frac{1}{4}x^4 + \frac{1}{2}ax^2 + bx$
swallowtail	$\frac{1}{5}x^5 + \frac{1}{3}ax^3 + \frac{1}{2}bx^2 + cx$
butterfly	$\frac{1}{6}x^6 + \frac{1}{4}ax^4 + \frac{1}{3}bx^3 + \frac{1}{2}cx^2 + dx$
hyperbolic umbilic	$x^3 + y^3 + ax + by + cxy$
elliptic umbilic	$x^3 - 3xy^2 + ax + by + c(x^2 + y^2)$
parabolic umbilic	$x^2y + y^4 + ax + by + cx^2 + dy^2.$

You can see from the list that there is no *obvious* pattern, no *obvious* reason why these seven alone should occur. In fact the proof of Thom's theorem makes essential use of some very deep results from multidimensional topology, analysis, and abstract algebra – as I said, mathematics is a harmonious whole – and even then the proof is very difficult.

The geometrical shapes of the elementary catastrophes are very beautiful. Figure 189 shows a computer-drawn cross-section of part of a parabolic umbilic.[5]

Figure 189

You will observe that Zeeman's machine corresponds to the cusp catastrophe on Thom's list. The same catastrophe illustrates how the theory applies to problems in biology.

A living cell is a 3-dimensional blob of matter. For clarity, I'll oversimplify and assume it is a 2-dimensional blob, which makes the pictures easier to draw. Correspondingly we will assume the cell lives in a control space of 2 dimensions.

We shall measure the behaviour of the cell by looking at the concentration of some fixed chemical. (It might be sodium chloride, it might be DNA: the principle is the same.) Since cells may undergo discontinuous changes – and it is exactly these that interest us – some catastrophe must be involved in the way the chemical concentration depends on the controls. A possible mode would be the cusp catastrophe.

As time passes, the chemical concentrations will gradually 'drift'; and we can represent this by moving the cell slowly through the control space. Figure 190 shows four stages in the development of the cell.

The position of the cell is shown in the lower half of the picture; the folded surface represents its chemical state. In the final picture there is a sharp line of discontinuity across the cell. The points to the left of this discontinuity are those which passed to the left of the fold point, and so have a high concentration of the chemical: the points to the right have a low concentration.

The cell has, in fact, *divided* into two different cells, because that is the only way that a sharp discontinuity in chemical concentrations can happen.

As I said, this is a crude and simplified model. Cell-division is not quite this simple. But it *is* a discontinuous process of the kind

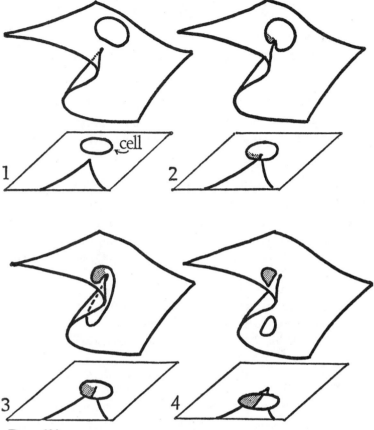

Figure 190

considered by Thom, so we expect each stage in it to be governed by one of the seven catastrophes.

This opens up a new kind of answer to the question, 'Why do cells divide?'. 'Because topological properties of their chemical state make it impossible for them *not* to divide.'

We glimpse a picture of the developing embryo: drifting gradually through some incredibly folded surface of chemical changes, dividing and subdividing; starting to form a limb here, a nerve, muscle, or bone cell there. And every step in the process occurs in just one of seven different ways.

Chapter 20   Foundations

An astronomer, a physicist, and a mathematician (it is said) were holidaying in Scotland. Glancing from the train window, they observed a black sheep in the middle of a field. 'How interesting,' observed the astronomer, 'all Scottish sheep are black!' To which the physicist responded, 'No, no! *Some* Scottish sheep are black!' The mathematician gazed heavenward in supplication, and then intoned, 'In Scotland there exists at least one field, containing at least one sheep, *at least one side of which is black*.'

Mathematicians (when on their best behaviour) tend to caution. A theorem *ought* to be true. The mathematician recalls the many occasions on which the 'obvious' has turned out to be wrong, and shudders. In a subject where regular 17-gons can be constructed but 19-gons cannot; where the sphere can be turned inside out;[1] where there are the same number of rational numbers as integers, who can blame him? And he decides that he must suspend judgement until the theorem is *proved*.

Not all mathematicians display such caution, I must add, least of all some of the world's greatest (alive or dead).[2] But even those who do not are usually aware that they are on dangerous ground. And it must be remarked that there is a great difference between suspending judgement on a theorem, and ignoring it. Anyone studying mathematics must be prepared to say 'I don't quite follow that bit, but for the moment I'll pretend I do, and see where it leads.' Often the difficulty is easier to understand in retrospect. A person who insists on understanding every tiny step before going on to the next is liable to concentrate so much on looking at his feet that he fails to realize he is walking in the wrong direction. It is always permissible to ignore difficulties the first time round; this way you can check up on the basic plan of attack. Then, if that seems all right, you can tidy up the details.

The time has now come to tidy up some details of our previous work. A sheep that is black on one side and white on the other is a considerable rarity; and on the whole it matters little whether or not a given sheep is exactly as it first appears. But mathematics has a distressing tendency to pile deductions on top of each other like a somewhat wobbly house of cards. Remove one card, and the whole structure topples. Early in the American space-programme, a rocket costing several millions of dollars had to be blown up just after take off. A semicolon had been omitted from the computer-tape controlling its guidance system. The more complex the structure, the more disastrous a flaw can be.

At the turn of the century mathematicians began to have doubts about the foundations of their subject. It is fashionable to talk about 'pyramidal' organizations. Mathematics resembles a pyramid stuck point downwards. Almost all of its results rest ultimately on a small number of assumptions. It is only common prudence to take a close look at those assumptions, and to make them as solid a base as possible.

A Half-Black Sheep in the Family

Frege realized that an adequate treatment of the concept 'number' was totally lacking. We mentioned his attempts to put numbers on a firm footing in Chapter 9. The keystone was the way we split up sets into classes, such that all the members of a given class were equinumerous. According to pragmatical attitude B, 'These classes behave like numbers, so I might as well say that they *are* numbers.'

In fact we did not adopt this attitude, preferring to take the existence of numbers as an axiom. This is fortunate, because taking the set of all sets with a certain property is less innocuous than it seems. Bertrand Russell pointed this out to Frege just as Frege had completed his masterpiece.

Imagine a librarian in a large library. Among the books on the shelves are certain catalogues: of poetry books, reference books, mathematics books, outsize books, ... Some of these (such as the catalogue of reference books) list themselves, others (poetry) do not. In order to clarify this situation the librarian decides to make a catalogue (call it *C*) of all the catalogues which do not list themselves.

The problem is: does catalogue *C* list itself?

If it does, then it is listed in *C*, so does not list itself. If it does not, then *C* is one of the catalogues which does not list itself, so must be listed in *C*.

Remember the village barber?[3]

If this were just a paradox about librarians we would not be in trouble: we could keep all references to librarians out of mathematics. But a *set* is much the same as a catalogue, and the things it lists are its members.

The set-theoretic version is this: let *B* be the set of all sets which are not members of themselves. Is *B* a member of itself, or not? The reasoning is just as for catalogues: whichever we assume, we deduce the other.

So set theory as used by Frege is *inconsistent*. No worse fate could befall any theory.

The only remedy is to abandon Frege's *naive* set theory and seek a replacement that is not inconsistent. In the naive theory we have allowed ourselves too much licence, and reaped the reward of the over-indulgent.

Two Remedies

In order to circumvent the Russell paradox, we have to change the rules so that the argument fails. But our new rules mustn't be too restrictive, or we may throw out the mathematical baby with the paradoxical bathwater.

There are at least two places in the argument where the logic is a tiny bit dubious.

In the first place, our freedom to construct sets may be too great. If *B* were not a set, then 'membership' of *B* might not make sense, so the argument could not be pushed through.

In the second, we may have placed undue reliance upon proof by contradiction. If not-not-*p* and *p* are *different* then proof by contradiction breaks down: all we have proved is that *B* is not a member of *B* and *not* not a member of *B*, and the latter does not contradict the former.

Advocates of the second school of thought – the so-called *intuitionists* – were particularly vocal in the 1930s. Their remedy is rather drastic: if we throw away proof by contradiction we lose an awful lot of mathematics. The intuitionists took great pains to reconstruct the major portions of mathematics without using proof by contradiction, and it is amazing how much can be saved; but nonetheless changes occur. All functions are continuous, for example.

The intuitionist argument runs along these lines: it is at first sight *plausible* that not-not-p is the same as p, or equivalently that exactly one of

$$p \qquad \text{not-}p$$

is true. Certainly if p refers only to a finite number of objects this should be the case, for we can in principle check p for each object. When we finish, either every object satisfies p, so p is true; or one of them doesn't and p is false, so not-p is true.

But if p refers to infinitely many objects, this option is no longer open. We may check as many objects as we like, and find p true – but we have no way of knowing whether, for an object still unchecked, p is false. Unless we can find a proof of p (or of not-p) that works for all objects, we are stuck. Now conceivably, p might be true of each object, but for a different reason in each case – a sort of infinite coincidence. *If* this happens, we certainly can't disprove p. But we can't prove p either, because we can't write out an infinitely long proof.

Consider, for example, the *Goldbach conjecture*: every even number greater than 2 is a sum of two primes. This has never been proved, or disproved. If you try it out for various even numbers it seems to work:

$$4 = 2+2 \qquad 18 = 5+13$$
$$6 = 3+3 \qquad 20 = 7+13$$
$$8 = 3+5 \qquad 22 = 3+19$$
$$10 = 3+7 \qquad 24 = 5+19$$
$$12 = 5+7 \qquad 26 = 3+23$$
$$14 = 3+11 \qquad 28 = 5+23$$
$$16 = 5+11 \qquad 30 = 7+23$$

On the other hand, no recognizable pattern occurs. It is certainly *possible* that no pattern exists, and yet the conjecture may be true.

This very possibility means that our confident assertion that either p or not-p is true is metaphysics, not mathematics. It is based on the assumption that infinitely many objects behave like finitely many objects. And we have seen enough examples of the curious behaviour of the infinite (especially Chapter 9) for this to be a questionable assumption.

If the assumption *is* wrong, then the Russell paradox would just be one of those theorems that we can neither prove nor disprove. Of course, just what the word 'wrong' may mean in this context is a matter for investigation – it's not clear that it has any useful meaning.

The intuitionist is happy with statements like 'Every even number less than 10^{100} is a sum of two primes', and he agrees that these are either true or false. But the statement 'Every even number is a sum of two primes,' might be neither true, nor false: it falls into a new category of truth – *dubious*.

Apparently less drastic is the idea that we should restrict our freedom to form sets. In one alternative set theory[4] there are two distinct types of set-like object. First there are *classes*; these have elements, and behave much like the naive sets. But a class is not necessarily able to be a member of another class. Those classes which *can* be members are the *sets*.

This means that a definition of a class in the form
$$C = \{x \mid x \text{ has property } P\}$$
must be interpreted as: C is the class of all *sets* x which have property P. If x has property P we cannot deduce that $x \in C$ *unless we know that x is a set.*

In the Russell paradox, the argument tries to show that if $B \notin B$ then B has the property which defines elements of B (namely, not being elements of themselves) and so lies in B. In our new set theory, this cannot be deduced unless B is a set.

Now we turn the Russell paradox on its head. All it does is *prove* (by contradiction) that B is not a set. For if B is a set the paradox works – contradiction.

Those classes which are not sets are called *proper classes*. The Russell paradox proves they exist. On the other hand, we don't know of any *sets*.

The only way to make sure that sets exist is to lay down axioms which say they do. Simple axioms, obviously necessary to any set theory, say that ∅ is a set, or that the union of two sets is a set, or that the intersection of two sets is a set. Thus we build up an *axiomatic* set theory.

Frege's naive set theory was modelled on the behaviour of collections of real objects. We don't expect the real world to contradict itself (a belief which may prove to be as unfounded as many another cherished tenet of humanity) and so we expect Frege's set theory to be consistent. It isn't, but in the final analysis this is because it has strayed beyond the realms of reality.

Axiomatic set theory, however, never gets anywhere near the real world. Before it can be an acceptable basis for mathematics, it should be demonstrated to be consistent. There is no reassurance on this point from the physical world. A *proof* of consistency is needed.

The Hilbert Programme

First we must decide which methods of proof are to be allowed in giving the demonstration of consistency. Clearly we cannot use methods of proof whose own consistency is open to doubt.

David Hilbert, who was the first to consider the question, felt that a satisfactory proof must be one whose techniques can be completely specified in such a way that, so to speak, a computer could carry them out. There must be no vagueness; each step must be perfectly clear, each contingency accounted for.

Hilbert also realized that *for the purpose of the proof* we must ignore any meaning that could be attached to the mathematical symbols. We should think of mathematics as a game played with symbols on paper, according to certain fixed rules. Rules saying, for example, that the combination of symbols $1+1$ may be replaced by the symbol 2. If we can show that however the game is played we cannot produce the combination of symbols

$$0 \neq 0$$

by a legal sequence of moves; and if we can prove this in a finite, constructive fashion; then we will have a proof of consistency.

If the combination $0 \neq 0$ *did* occur we could *interpret* the moves of the game as a proof that $0 \neq 0$, so axiomatic set theory would be inconsistent. On the other hand, if axiomatic set theory is inconsistent there is a proof that $0 \neq 0$, which gives a sequence of moves in the game.

As well as suggesting this, Hilbert laid down a complete programme for carrying out the proof. In order to give mathematics as sound a logical foundation as could be desired, it was only necessary to carry out the programme.

Hilbert was also interested in another problem: whether, in principle, every problem can be solved. This is connected with the intuitionist belief that some cannot. Hilbert's programme included an answer to this, too: he wished to show that there existed a definite procedure whereby one could decide in advance whether or not a problem could be solved. He was convinced that this was possible.

Hilbert at this time was the acknowledged leader of the mathematical world. But a young man, Kurt Gödel, who had trained as an engineer, was convinced that Hilbert was wrong. In 1930 he sent for publication a paper[5] which left the Hilbert programme in ruins. Another great mathematician, Von Neumann, was giving a series of lectures on the Hilbert programme. But when he read Gödel's paper he cancelled what remained of the course, in order to lecture on Gödel's work.

Gödel proved two things:

(i) If axiomatic set theory is consistent, there exist theorems which can neither be proved nor disproved.

(ii) There is no constructive procedure which will prove axiomatic set theory to be consistent.

The first result shows that problems are not always soluble, even in principle; the second wrecks Hilbert's programme for proving consistency. It is said that when Hilbert heard of Gödel's work he was 'very angry'.

Later developments have shown the wreck to be greater than even Gödel imagined. *Any* axiomatic system sufficiently extensive to allow the formulation of arithmetic will suffer the same defects. It is not any particular axiomatization that is at fault, but arithmetic itself!

Gödel Numbers

This and the next section provide an outline proof of Gödel's theorem. Anyone who wishes to omit these two sections may do so without losing the thread.

We begin with a simple question: how many arithmetical formulae are there? (By 'arithmetical formula' I mean any combination of the symbols $+, -, \times, \div, (,), =, 0, 1, 2, 3, 4, 5, 6, 7, 8, 9$.)

Obviously there are infinitely many. But bearing in mind Chapter 9, we ask: which infinity? Countable or uncountable? In fact, the set of formulae is countable. There exists a bijection between it and the set **N** of whole numbers.

We begin by 'coding' the symbols:

$+$	$-$	$\times$	$\div$	$($	$)$	$=$	0	1	2	3	4	5	6	7	8	9
1	2	3	4	5	6	7	8	9	10	11	12	13	14	15	16	17

Now, to code a *string* of symbols, such as

$$4+7 = 11$$

we form the number

$$2^{12}.3^1.5^{15}.7^7.11^9.13^9$$

where $2, 3, 5, 7, 11, 13, \ldots$ is the sequence of primes, and the powers $12, 1, 15, 7, 9, 9$ are the codes of the symbols $4, +, 7, =, 1, 1$ of the string.

In this way we can associate with each string a code, which will be a whole number.

Because of the uniqueness of prime factorization, we can reconstruct a string from its code. Thus, if the code is 720, we factorize

$$720 = 2^4.3^2.5^1.$$

The symbols whose codes are 4, 2, 1 are $\div, -, +$; so 720 is the code of the string

$$\div - +$$

(not a very meaningful string, but a string nevertheless.)

For more complicated strings, the numbers get very large. But every string has a code, and different strings have different codes. By arranging the strings in order of size of their codes, we can see that the set of strings is countable.

In axiomatic set theory we have more symbols: $\in$, $\cup$, $\cap$, {, }, and also symbols for 'variables' $x, y, z, \ldots$. But the same argument applies: code the symbols, then code strings using prime numbers. So in axiomatic set theory the set of strings is also countable.

Proof of Gödel's Theorems

For this section we shall be working in two different systems: an axiomatic set theory $\mathscr{S}$, and ordinary arithmetic, $\mathscr{A}$. The system $\mathscr{S}$ will be a formalization of arithmetic. Inside $\mathscr{S}$ we will have certain symbols, from which we construct strings; the axioms of $\mathscr{S}$ tell us what we are allowed to do to the strings.

We assume that $\mathscr{S}$ is so arranged that the arithmetical symbols are used in $\mathscr{S}$ with their usual meanings; so that the string $2+2 = 4$ has two interpretations: (i) a string in $\mathscr{S}$, devoid of any meaning, (ii) a formula in arithmetic. Further, if a sequence of permissible changes of strings in $\mathscr{S}$ leads us (say) to the string $2+2 = 4$, then the corresponding sequence of arithmetical formulae will be a proof, in $\mathscr{A}$, that $2+2 = 4$.

In $\mathscr{S}$, certain strings will involve a single numerical variable x. Examples are the strings

$$x+1 = 1+x$$
$$x(x-1) = xx-x \qquad (\dagger)$$
$$x+x = 43.$$

We are particularly interested in this kind of string: to save breath we call a string involving the numerical variable x a *sign*.

If a is a sign and t is a positive number, we can form a new string $[a:t]$ by substituting t for x in a. (Here, of course, t is thought of as a string of symbols $0, 1, 2, 3, \ldots$.) For example, if a is the sign $x+1 = 1+x$ and $t = 31$, then $[a:31]$ is the string $31+1 = 1+31$.

Every sign has a Gödel number. We arrange these in order, and let

$$R(n)$$

be the nth sign. Then every sign is equal to some $R(n)$ for suitable choice of n.

Now define a set K of whole numbers (in $\mathscr{A}$) by: $n \in K$ if and only if $[R(n):n]$ is not provable in $\mathscr{S}$.

For example, to see if $3 \in K$ we find $R(3)$: say it is the string $x+4 = 0$. Substituting 3 for x we get the string $3+4 = 0$. If this is not provable in $\mathscr{S}$, then $3 \in K$.

Now the formula $x \in K$ in $\mathscr{A}$ can be formalized in $\mathscr{S}$, and gives some string S in $\mathscr{S}$. Now S involves a single numerical variable, so is a sign. For any particular number n the string $[S:n]$ is a formal version of the arithmetical statement $n \in K$.

Since S is a sign we have $S = R(q)$ for some q. Now we show that the string

$$[R(q):q] \tag{1}$$

is not provable in $\mathscr{S}$, but that at the same time

$$\text{not-}[R(q):q] \tag{2}$$

is also not provable in $\mathscr{S}$.

If (1) can be proved in $\mathscr{S}$ then its interpretation in $\mathscr{A}$ is true, $\mathscr{S}$ being a formalization of $\mathscr{A}$. So $q \in K$. But by the definition of K, it follows that (1) is *not* provable in $\mathscr{S}$.

If (2) is provable in $\mathscr{S}$, then not-$(q \in K)$ is true in $\mathscr{A}$. So $q \notin K$, whence it is false that $[R(q):q]$ is not provable in $\mathscr{S}$; so $[R(q):q]$ is provable in $\mathscr{S}$. Assuming $\mathscr{S}$ consistent, it follows that (2) is *not* provable in $\mathscr{S}$.

The string $[R(q):q]$ (which is a perfectly definite string in $\mathscr{S}$) therefore gives a theorem which can neither be proved nor disproved in $\mathscr{S}$. *This proves Gödel's first theorem.*

Disentangling, you will find that $[R(q):q]$ can be interpreted as asserting its *own* unprovability. It says, almost, 'This theorem is unprovable,' which is very like, 'This sentence is false.' However, 'This theorem is unprovable,' cannot be formalized in $\mathscr{S}$, which is why we have to hop around from $\mathscr{S}$ to $\mathscr{A}$ and back again.

Now we can prove Gödel's second theorem. Let T be the string $[R(q):q]$, which we have seen asserts its own unprovability. Let W be any formula in $\mathscr{S}$ which asserts the consistency of $\mathscr{S}$. We want to show that W cannot be proved in $\mathscr{S}$.

Gödel's first theorem reads 'if $\mathscr{S}$ is consistent, then T is not provable in $\mathscr{S}$'. We can express this in $\mathscr{S}$. '$\mathscr{S}$ is consistent' is our formula W; 'T is not provable in $\mathscr{S}$' is just T itself, because T

asserts its own unprovability. So Gödel's first theorem, written in $\mathscr{S}$, takes the form

$$W \text{ implies } T.$$

If we could prove W in $\mathscr{S}$, then this would enable us to prove T. But we know that T cannot be proved; hence W cannot. Since W asserts the consistency of $\mathscr{S}$, it is not possible to prove S consistent within $\mathscr{S}$. *This is Gödel's second theorem.*

Undecidability

In full detail (which involves specifying carefully what is meant by a 'constructive procedure') Gödel's theorems can be given a completely watertight proof.[6] Although (2) deals the death-blow to the Hilbert programme, it is (1) that is more interesting. It shows that in ordinary arithmetic there exist statements P such that neither P nor not-P can be proved. Such statements are said to be *undecidable*.

In a way, this is a partial vindication of intuitionism; but only if one equates 'provable' with 'true'. And the proof of Gödel's theorem applies equally well to intuitionist mathematics.

Several questions raised by Hilbert have gone the same way as the problem of arithmetical consistency. A *Diophantine equation* is a polynomial equation, such as

$$x^2 + y^2 = z^3 t^3$$

for which we seek solutions in *integers*. Hilbert asked for a method of testing a given Diophantine equation, to see whether it had a solution. Matijasevič (following earlier work by Davis, Putnam, and Robinson) has recently proved[7] that no such method exists: whether or not a given Diophantine equation has solutions may be undecidable.

An amazing corollary of Matijasevič's proof is that there exists a polynomial

$$p(x_1, x_2, \ldots, x_{23})$$

in 23 variables, whose *positive* values for integer values of the variables run through exactly all prime numbers. A 'formula for primes', no less![8] In principle, it would be possible to write the polynomial down explicitly; but in practice it is too complicated to do more than give a procedure whereby it *could* be written down. And it is unlikely to be of any use in the theory of prime numbers.*

In Chapter 9 we mentioned this problem: is the cardinal c of the real numbers the next cardinal after $\aleph_0$? This is the *Continuum hypothesis*. Hilbert asked whether it was true or false (although Cantor was the first to ask the question). Cohen's answer,[9] in 1963, was 'Yes and no'. It is *independent* of the other axioms of set theory. You can add an axiom saying that the Continuum hypothesis is true, and you will not make set theory inconsistent (assuming it was consistent to start with!); or you can add an axiom saying that it is false, and again no inconsistency will arise. This is a twentieth-century version of non-Euclidean geometry: by denying the Continuum hypothesis we can produce non-Cantorian set theories.

Epilogue

Perhaps it should have been clear from the start that the Hilbert programme could not succeed. It is too like trying to lift oneself by one's bootstraps. Can *any* knowledge be true in an absolute sense? But the value of Gödel's work is that it goes beyond mere philosophical speculation: it *proves* the impossibility of an arithmetical proof of the consistency of arithmetic.

This does not mean that we cannot find other ways of proving arithmetic consistent. Gentzen[10] has indeed proved this; but his methods involve *transfinite induction* – I won't go into what that is – and of course the consistency of *that* is open to doubt.

So the foundations of mathematics remain wobbly, despite all efforts to consolidate them. Perhaps one day somebody will find an inescapable contradiction, and the whole subject will collapse. But even then, there would be indefatigable mathematicians pottering around in the ruins, tinkering with the works, and trying to resurrect what they could.

For the truth is that intuition will always prevail over mere

* But see Appendix, p. 305.

logic. If the theorems fit together properly, if they yield insight and wonder, no one is going to throw them away just because of a few logical quibbles. There is always the feeling that the *logic* can be changed; we would prefer not to change the theorems.

Gauss called mathematics 'the queen of the sciences'. I prefer to think of it as an emperor. And though it may yet transpire that the emperor has no clothes, he is still better dressed than his courtiers.

Appendix

And still it moves . . .

'There is probably no other science which presents such different appearances to one who cultivates it and one who does not, as mathematics. To [the latter] it is ancient, venerable, and complete; a body of dry, irrefutable, unambiguous reasoning. To the mathematician, on the other hand, his science is yet in the purple bloom of vigorous youth' – C. H. Chapman, in 1892

The reader who has persevered this far must by now be a cultivator of mathematics, even if he was not at the start of the endeavour. He will therefore appreciate that, while it may be ancient and venerable, it is far from complete; that not all of it is dry; and that its reasoning has not always been either unambiguous or irrefutable – nor is yet. Purple bloom and youthful vigour, however, may appear a trifle far-fetched; in any case, what was true in 1892 may not be today. My remark in Chapter 1 that most 'modern mathematics', as taught in schools, is over a century old, points to one possibility: if this truly *were* modern mathematics, one might conclude that the well had run dry. Fortunately it has not: the present century has witnessed the most rapid and broad advances of the subject in the history of humanity, as indeed it has in many other subjects.

So youthful is the vigour, indeed, that it has caused what might have been a minor embarrassment to your humble narrator. The first twenty chapters of this book were written in 1973–4 and published in 1975. It is now 1980. Even by 1977 several discoveries had been made which necessitated changes to those chapters. It would have been possible to do this by rewriting parts of them but it seemed better to make a virtue out of necessity, let what I wrote stand except for a few pointers toward this chapter, and add an appendix to update the information and, at the same time, demonstrate how rapidly mathematics is evolving.

It is neither feasible nor desirable to survey everything that has happened in those few years; just to catalogue the results would take a good many books the size of this one. My selection is strongly conditioned by links with the material discussed above. It cannot, therefore, be truly representative of the great variety of activities in which mathematicians indulge.

In Chapter 11 I unguardedly remarked, of the four-colour problem, that '. . . it would be a pity if it *were* solved . . .': you might think that a problem that was posed in 1852 and unsolved in 1975 would have the decency to *remain* unsolved a little longer. It would seem, however, that the Principle of Universal Cussedness of Things (known by other, less flattering names) is in operation, for in mid 1976 a solution to the problem was announced. Although it is not quite certain that the proof is in the bag, for technical reasons that I shall mention, it looks as if it is. The next section of this chapter tells the story to date. Following that, we review recent work on polynomial formulae for primes, which among other things disproves something I said in Chapter 20. Finally, I want to expand on a point touched upon in Chapter 14; namely, the use of topology to study dynamical systems. Here some topological gadgets known as *strange attractors* have suddenly begun to manifest themselves in ecology, geology, and fluid dynamics, as a phenomenon that the ecologists have christened *chaos*. One result is that the line between deterministic and random behaviour cannot be drawn as clearly as has hitherto been widely assumed.

The Four-Colour Theorem

In Chapter 11 we had a four-colour problem and a five-colour theorem. On 22 July 1972 two mathematicians at the University of Illinois, Kenneth Appel and Wolfgang Haken, announced a proof[1] that every map on the plane can be coloured with *four* colours, resolving the long-standing question. The snag with their proof is that it involves an immense amount of computing time – they used about 1200 hours, but a check might be performed in 300 on a big machine – and a single error might destroy the proof completely. Contrary to popular opinion, computers *do* make errors even when their programmers do not. The problems raised by this are as much philosophical as technical (is a proof checked by a human any more likely to be accurate than one checked by computer?) and look like causing some problems in future. At the present moment, the Appel–Haken proof has not been fully checked by an independent program, although those parts that have been checked appear to be error-free. Philosophy apart, most mathematicians would probably agree that the theorem is *proved* provided such a check is carried out in full. The philosophical problem is an interesting one, however, because it involves the difference between a *logical* proof and a *satisfactory* one. I shall return to this once we've seen how it goes.

The proof of the five-colour theorem given in Chapter 11 is essentially due to A. B. Kempe (1879). Let us recall it briefly. It makes use of a *reduction process* (which involves a number of different cases) to show that the given map may be coloured provided another map, with *fewer* regions, can be coloured. By applying this process over and over again we must, since the number of regions decreases at each stage, eventually reach a map with as many or fewer regions than we have colours. *This* map is manifestly colourable, therefore reversing the reduction step by step, so was the previous one, and the one before that, and . . . the original.

Kempe went further than we have done (arguing for five colours): he added an argument purporting to prove that *four* would suffice (involving extra cases in the reduction). The mathematicians of the time seem to have accepted this 'proof' without a qualm, until in 1890 P. J. Heawood pointed out the mistake. The proof we have given above, as Heawood also pointed out, remains valid for *five*-colourability.

The basic strategy of our proof for five colours and the Appel–Haken proof – and even Kempe's fallacious attempt – is the same. (Kempe simply failed to carry it out correctly.) For our present purposes this emerges more clearly if we recast our proof in a slightly different form. The idea is to argue by contradiction and suppose that there *does* exist a map requiring more than five colours. There is then such a map having the *least* possible number of regions (in trade jargon, a *minimal criminal*), say M. By definition, every map with fewer regions than M can be coloured in five colours.

Next we show that M *must* contain at least one of a list of configurations, namely Figures 109, 110, 111, 112. Then we consider one of these configurations and concoct an argument of the following kind. From any map M which contains it, we show how to derive another map M' with fewer regions. This M' is so constructed that, *if* it is five-colourable, *so is* M. Thus the burden of colourability shifts from M to the simpler map M'. (In our proof on pages 169–73, M' is obtained from M by merging regions, as indicated.) We concoct such arguments (which may differ in detail) for each configuration in our list.

What happens when we apply these ideas to a minimal criminal? By definition, a minimal criminal has a very strong property: *every* map with fewer regions is five-colourable. It follows that, whichever configuration on our list is encountered, the *smaller* map M' is five-colourable. However, the general argument which shifted the burden to M' now tells us that M *itself* is five-colourable. This argument works provided one of our listed configurations is known to occur in M, but we have cunningly chosen our list so that *every* map contains at least one configuration! Hence M contains at least one and *we apply the argument to that one*. Therefore M is five-colourable.

This, however, contradicts the criminality of M!

More carefully: we started by assuming M was *not* five-colourable; now we have proved that it is. It follows, just as on p. 116, that *our initial assumption must be wrong*. That is, M is not a minimal criminal: so no minimal criminals exist. Hence *no criminals at all exist*. (Big crime cannot exist without small crime, or, more precisely, if any criminals exist then the smallest one is also minimal.) Therefore every map is five-colourable.

To grasp this argument 'in a nutshell' we introduce two new terms. We say that a set of configurations is *unavoidable* if every map contains at least one configuration in the set. A configuration is *reducible* if it cannot be contained in a minimal criminal. The proof works by constructing an unavoidable set of reducible configurations.

We can try the same thing on the four-colour problem (noting that the condition for reducibility is now that the configuration is not contained in any minimal criminal for *four*-colourability); and this is what Kempe tried to do. He made a mistake (in a complicated argument about rearranging colours) when considering Figure 112. Appel and Haken reasoned that the way to get round this is to replace Kempe's four possibilities by a different unavoidable set, by throwing away the configuration that caused him trouble. It is of course necessary to add *several* new configurations, because the possibilities increase rapidly as the size of the configurations considered grows. Having found a new unavoidable set, it is tested for reducibility. *If this doesn't work, we throw away the bad sets and try again with even more.* You can see the danger: the attempted proof works by chasing its own tail and succeeds only if it catches it! Now Appel and Haken developed such a good 'feel' for the configurations needed that they managed to come up with something that *did* catch its own tail (at least, if the computations are correct). Unfortunately, their unavoidable set contained 1936 configurations! It has now been reduced to 1877, but it is doubtful if it can be made very much smaller.

So good was their 'feel' that, after using the computer for several years to explore the problem, they actually wrote down the unavoidable set *by hand*. The computer comes in to check unavoidability and, more seriously, reducibility. There is a lot of interesting mathematics involved in how this is done, but I won't go into details. For moderately large configurations, reducibility takes a lot of checking, and it is here that the computer time and the danger of error are greatest.

One might expect that the solution of such a notorious unsolved problem would have caused a great stir in the mathematical world. To some extent it did, but the excitement was mostly short-lived and the overall impact a good deal less than that of solutions to other problems of mathematical importance but less familiar to the man in the street. I don't mean to belittle Appel and Haken's achievement, which is a remarkable *tour de force* whose influence will be felt for a long time, but I think the description is a fairly objective one. There are several reasons why the solution was a bit of a damp squib.

The most immediate is that, while the four-colour problem is something that every mathematician knows about, and may occasionally make a half-hearted attempt to solve, it is not something that many of them *worry* about. The reason is that, as far as the mainstream of mathematics is concerned, it doesn't seem to matter much whether the answer is four or five. Nothing of importance hinges upon it. It is not clear that, if you know the answer, you can *do* anything with it. There are no powerful techniques, of general use in mathematics, that would be advanced if the result were known. Colourability is a complete side issue in topology, let alone the rest of mathematics. Having said that, I should add that this does not diminish the *difficulty* of solving the problem: there can be few mathematicians who would not have been immensely pleased with themselves had they been able to match Appel and Haken's achievement.

Secondly, there was the problem of whether the computerized proof was correct. It's hard to get excited about something when you are not certain it has been done! For many years past there have been rumoured solutions, all with mistakes, and this breeds a certain scepticism.

The third reason, and in many ways the most serious, was the *nature* of the proof. In a sense which I shall have to elucidate, it doesn't give a satisfactory explanation *why* the theorem is true. This is partly because the proof is so long that it is hard to grasp (including the computer calculations, impossible!), but mostly because it is so apparently structureless. The answer appears as a kind of monstrous coincidence. *Why* is there an unavoidable set of reducible configurations? The best answer at the present time is: there just is. The proof: here it is, see for yourself. The mathematician's search for hidden structure, his pattern-binding urge, is frustrated.

You see, there is more to mathematics than just logical rigour. The proof of Pythagoras' theorem in Chapter 2 Figure 1 is not even logically rigorous without a lot of extra work on making precise the concepts involved; but it is immediately *convincing*. The truth of the result – the *necessity* for it to be true – impresses itself upon the mind. Likewise, the proof of the five-colour theorem can be grasped in its entirety. Even very long proofs, or whole areas of mathematics, can be grasped in this way. A *satisfactory* proof is one that can communicate how a theorem sits within the overall scheme of things. It is probably true that Appel and Haken have such a grasp of the four-colour problem – I doubt they could have constructed their unavoidable set without it – but their proof does not communicate this understanding to anyone else.

There are at least two possibilities here. One is that a different, more structured proof will now be found. Often this happens: it is much easier to prove a theorem once you know it is true. (For a start, the effort you put in is less likely to be wasted.) The other, which Appel and Haken themselves consider more likely, is that the nature of the proof reflects the nature of the problem. It *is* a monstrous coincidence. They further conjecture that the future will reveal a good many more theorems of the same type.

Now, whether the four-colour theorem itself is of this type I am unsure, though it certainly looks that way, but it is extremely likely that Appel and Haken are right when they suggest that such theorems exist. We have seen that in Hilbert's day the prevailing belief among mathematicians was that every true theorem must have a proof, and how naïve Gödel showed this belief to be. It is probably just as naïve to imagine that every provable theorem has an intellectually satisfying proof.

Polynomials and primes

In Chapter 20 I mentioned Matijasevič's prime-representing polynomial, saying: 'In principle, it would be possible to write the polynomial down explicitly; but in practice it is too complicated to do more than give a procedure whereby it *could* be written down. And it is unlikely to be of any use in the theory of prime numbers.' Strictly, this statement remains true of the actual polynomial found by Matijasevič, but the impression it gives (which is what I thought at the time) is that *any* prime-representing polynomial would be too complicated to write down explicitly.

I was wrong.

I was also arguably wrong on the utility of such formulae in the theory of primes – though it remains true that it is seldom the best starting point, when discussing primes, to use a polynomial formula – in the sense that there *are* properties of primes which follow from such formulae and were not known previously.

To whet your appetite, here is a polynomial formula for primes, due to J. P. Jones, D. Sato, H. Wada, and D. Wiens[2]. *The set of primes is identical with the set of positive values of the polynomial*

$$(k+2)\{1-[wz+h+j-q]^2-[(gk+2g+k+1)\cdot(h+j)+h-z]^2$$
$$-[2n+p+q+z-e]^2-[16(k+1)^3\cdot(k+2)\cdot(n+1)^2+1-f^2]^2$$
$$-[e^3\cdot(e+2)(a+1)^2+1-o^2]^2-[(a^2-1)y^2+1-x^2]^2-[16r^2y^4(a^2$$
$$-1)+1-u^2]^2-[((a+u^2(u^2-a))^2-1)\cdot(n+4dy)^2+1-(x+cu)^2]^2$$
$$-[n+l+v-y]^2-[(a^2-1)l^2+1-m^2]^2-[ai+k+1-l-i]^2-[p$$
$$+l(a-n-1)+b(2an+2a-n^2-2n-2)-m]^2-[q+y(a-p-1)$$
$$+s(2ap+2a-p^2-2p-2)-x]^2-[z+pl(a-p)+t(2ap-p^2-1)$$
$$-pm]^2\}$$

as the variables a,b,c,d,e,f,g,h,i,j,k,l,m,n,o,p,q,r,s,t,u,v,w,x,y,z range over the natural numbers 0,1,2,3, . . .

The inventors point out: 'Note the apparent paradox. The polynomial factorizes!' Namely, it has factors $k+2$ and the monster inside the curly brackets. How, then, can it represent primes, which *don't* factorize?

The secret is that the monster expression is of the form $1-M$, where M is a *sum of squares*; hence $M \geq 0$ no matter what values we substitute for the variables. We construct M to have the property

$k+2$ is prime if and only if $M(k,$ other variables$) = 0$.   (†)

Then $(k+2)(1-M)$ is positive if and only if $1-M$ is positive, which entails $M = 0$. Thus the factorization becomes $(k+2)(1)$ where $k+2$ is prime by (†). Demise of paradox.

I haven't the space to give the proof[2] and in any case this is extremely technical. Some applications and further discussion are to be found in a beautiful paper by M. Davis, Yu.Matijasevič, and J. Robinson.[3]

Chaos

One rapidly growing branch of modern mathematics, the potential applications of which are only just beginning to be realized, is the topological theory of dynamical systems, whose guiding force is the American mathematician Stephen Smale (mentioned in Chapter 14 in connection with the Poincaré conjecture). The mathematical part of René Thom's 'catastrophe theory'[4] is essentially a part of dynamical systems theory: indeed, the starting point for Thom's ideas *is* dynamical systems theory, even if his applications and speculations are not.

Thom's 'elementary catastrophes' (Chapter 19) are the simplest examples of what dynamical systems theorists call *bifurcation*, but there are more complicated types, which Thom dubs 'generalized catastrophes'. Some of these have recently turned up in connection with population models and there are signs that more applications are on the way. Therefore, partly to demonstrate that the elementary catastrophes are not the only thing around and partly for their intrinsic interest, I would like to discuss the ideas at some length.

Consider a (idealized) population of rabbits. If at time t we have x rabbits, then at time $t+1$ we have $5x$ rabbits. Starting with 100 rabbits, how many do we have at time T? The pattern is pretty clear:

> at $t = 0$ we have 100 rabbits,
> at $t = 1$ we have $5.100 = 500$ rabbits,
> at $t = 2$ we have $5.5.100 = 2500$ rabbits,
> at $t = 3$ we have $5.5.5.100 = 12500$ rabbits,
> . . .
> at $t = T$ we have $5 \ldots 5.100 = 100.5^T$ rabbits.

This is the famed *exponential growth* often observed in populations (from bacteria to humans). In all cases it becomes unrealistic for large T: the universe isn't big enough. For our rabbits, for example, a rough estimate[4] shows that after 114 generations the total volume of rabbit exceeds that of the observable universe. Of course the formula breaks down far sooner: the number of rabbits begins to outstrip the food supply and the birthrate drops.

Instead of exponential growth, the numbers start to flatten out (saturation) or even to decline.

Now we can express the rule that led to exponential growth as follows. Let x_t be the number of rabbits at time t. Then

$$x_0 = 100$$

and

$$x_{t+1} = 5x_t.$$

This last equation is called a *recurrence relation*, because it expresses the value at time $t+1$ in terms of that at previous time t.

To take account of saturation effects, ecologists have introduced a variety of modifications to this equation. In all of them the idea is to reduce the growth rate as x_t becomes large. A favourite approach leads to the equation

$$x_{t+1} = kx_t(1-x_t).$$

Here we choose to measure population x_t in units making $0 \leq x_t \leq 1$. Then k is a constant and

$$0 \leq k \leq 4,$$

because $k > 4$ makes it possible to have $x_{t+1} > 1$ when $x_t = 0 \cdot 5$, and we have chosen our units to rule this out.

The constant k gives a measure of the 'unrestrained' birthrate, but the factor $1-x_t$ leads to a cut-off in growth as x_t nears 1. This, too, is a recurrence relation, but a *nonlinear* one, and it does not have any nice formula for the solution x_t in terms of t. To begin to understand it, we can try calculating some numerical values. A useful way to represent the results is to draw a *time-series graph*, plotting x_t against t. Figure 191 (a and b) shows several such graphs for various values of k.

We observe that for $k = 0 \cdot 9999$, the population rapidly dies out. For $k = 1 \cdot 5$ it settles down steadily to a value which, on the graph, seems to be close to $0 \cdot 33$. For $k = 2 \cdot 5$ it again settles down, but oscillates on the way. For $k = 3 \cdot 05$ it settles down into a pattern of steady oscillation between about $0 \cdot 5$ and $0 \cdot 7$. For $k = 3 \cdot 5$ there is a similar, less steady, oscillation, with a smaller variation occurring, in cycles of length 4. At $k = 3 \cdot 565$ the oscillations are much wilder, even more so at $k = 3 \cdot 7$, while by $k = 4$ all trace of pattern seems to have vanished (Figure 191c).

Readers who have access to a pocket calculator (the simplest kind will suffice) are encouraged to secrete themselves in a corner

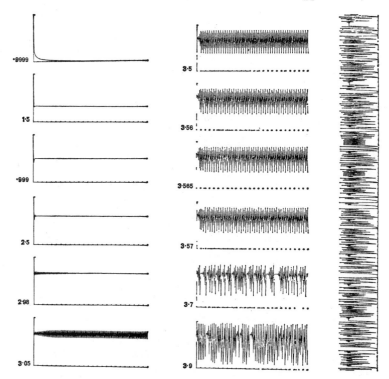

Figure 191a, b, c

somewhere and work out values of x_t for longer ranges of t, with different starting points and different values of k. They will find that Figure 191 is representative of the general behaviour of this recurrence relation. The wild behaviour of the later cases, with k near 4, continues indefinitely: the system never seems to settle down to any steady behaviour at all. This is the ecologist's *chaos*.

Now experiment on pocket calculators is all very well, but it neither *proves* that this sort of behaviour will occur nor does it *explain* why. It is here that the dynamical systems approach makes itself felt.

We are in fact dealing with what topologists call a *discrete* dynamical system – as opposed to a continuous one – and this simplifies the exposition while making the name more obscure, for we may define a discrete dynamical system as a *continuous* function $f: X \to X$ where X is a topological space. Put this way, all we have done is introduce a fancy name for something we already know about. But the change of name introduces a change of emphasis, because what we are interested in is *what happens if we do f repeatedly*. Roughly, f is how the system moves in one unit of time and we want to deduce its behaviour after many units of time.

For instance – and this is perhaps too simple an example to be entirely representative – suppose that X is a circular disc and f is 'rotate by 1 degree'. Then performing f n times in succession rotates the disc by a series of 1-degree clicks through a total angle of n degrees. You can think of this process as a kind of discrete analogue of a continuous rotation which at time t has rotated the disc through an angle of t degrees.

In general we define the *iterates* $f^{(n)}$ of f by

$$f^{(2)}(x) = f(f(x))$$
$$f^{(3)}(x) = f(f(f(x))),$$

and so on. Then the 'path' followed by a point x of X is the sequence of points

$$x, f(x), f^{(2)}(x), f^{(3)}(x), \ldots, f^{(n)}(x), \ldots$$

This sequence of points is called the *orbit* of x. To find out where x goes to after a long period of time we follow along the orbit and attempt to describe the way it jumps around in the space.

Some rather special but interesting things can happen. The most striking is if x is a *fixed point*, that is,

$$f(x) = x.$$

For a rotating disc as above, the only fixed point is the centre of the disc. For $k = 1 \cdot 5$ in our nonlinear recurrence relation, $x = \frac{1}{3}$ is a fixed point, since $1 \cdot 5 \times \frac{1}{3} \times (1 - \frac{1}{3}) = \frac{1}{3}$.

Central to the discussion that follows is a related notion, the *periodic point*. This is a point x which returns to its initial position after some number, n, of iterations of f, and hence thereafter repeats the same sequence of moves in a cycle of length n. In symbols, x is *periodic* with *period n* if

$$f^{(n)}(x) = x.$$

This is equivalent to being a fixed point of $f^{(n)}$, of course.

For example, with our rotating disc, *every* point x is periodic with period 360, because $f^{(360)}$ rotates through 360 degrees, bringing everything full circle. This is decidedly unusual and 'most' discrete dynamical systems have periodic points occurring less often!

The orbit of a periodic point (of period n) is a set of n points (fewer if some smaller $f^{(m)}(x) = x$, as can occur: the difference is immaterial for present purposes). It is called an *attractor* if the orbits of all points sufficiently close to it approach it more and more as time passes, and a *repellor* if they move further away. (The precise notions require more careful definition, but the examples that follow make it clear what is involved.) The importance of a periodic attractor is that all points near it behave in approximately periodic fashion, so that their behaviour can be described to high accuracy for arbitrarily long periods of time: we 'know what they will do in the future'. Repellors have just the opposite significance: we rapidly lose track of any nearby points not actually *on* the periodic orbit. However, the two tend to be associated and it is methematically important to consider both.

We now have most of the language needed to investigate our ecological population model. To set it up as a dynamical system let X be the unit interval
$$X = \{x \in R | 0 \le x \le 1\}$$
and define
$$f{:}X \to X, \quad f(x) = kx(1-x).$$
This links up with our earlier recurrence relation as follows: under iteration,
$$x_n = f^{(n)}(x_0).$$
So the behaviour of x_0 under iteration is given by its orbit under f.

The topological analysis[5] proceeds by examining the *graph* of f and its iterates. In general, the graph of f has the following shape:

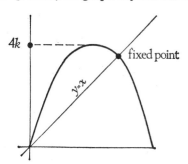

Figure 192

The fixed points of *f* can be read off as the values of *x* for which the graph of *f* meets the diagonal line $y = x$, as shown in the figure. Further, there is a vivid graphical method for iterating *f*, which I will illustrate as we go.

First suppose $k < 1$. Then the graph of *f* never rises above the diagonal and the only fixed point is 0. Iterates of a given x_0 can be found by drawing a 'staircase' which bounces off the diagonal as shown. (The bounce transfers the *y*-value $f(x)$ to the *x*-axis, ready to apply *f* to it again.)

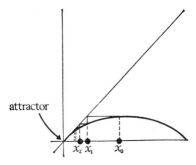

Figure 193

Clearly, whatever starting point x_0 we take, the orbit rapidly homes in along the staircase to the origin, which is therefore an *attractor*, indeed the *only* attractor. Its *basin of attraction* – the set of points that home into it – is the whole of *X*. This explains why, for $k < 1$, the population 'dies out': compare Figure 191 with $k = 0.9999$.

If $1 < k < 2$, then a new fixed point appears and it is an *attractor*. The origin becomes a repellor.

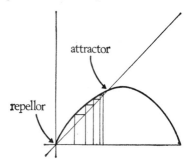

Figure 194

Further, the orbit approaches the attractor in a *steady* way, getting closer and closer but not oscillating around it. This compares with $k = 1.5$ in Figure 191. (We have already seen that $x = \frac{1}{3}$ is a fixed point in this case.)

For $2 < k < 3$ the picture is similar, except that the 'staircase' becomes a spider's web, spiralling around the intersection point.

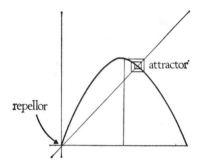

Figure 195

The fixed point remains an attractor but now the approach to it oscillates from side to side. This is just what happens in Figure 191, $k = 2.5$.

For $k > 3$, however, the spider's web spirals *out*wards, away from the intersection point. The fixed point becomes a *repellor* and the other fixed point at the origin remains a repellor, too.

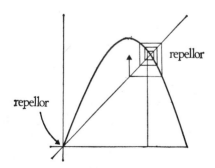

Figure 196

(By using calculus we can see why this happens. The curve
$$y = 3x(1-x)$$
crosses the diagonal line
$$y = x$$
at $x = 2/3$. At this point its gradient is given by
$$\frac{dy}{dx} = 3-6x = -1,$$
which means that it crosses the diagonal at right angles. For $k < 3$ the angle is less, for $k > 3$ it is more. This difference in the angle determines the direction of spiral.)

It follows that the system does not, in general, settle down towards a fixed point. Where *does* it go?

To find out, we look at the second iterate $f^{(2)}$. A simple calculation gives
$$f^{(2)}(x) = f(f(x)) = k(kx(1-x))(1-kx(1-x))$$
$$= k^2x - (k^2+k^3)x^2 + 2k^3x^3 - k^3x^4.$$
Figure 197a shows the graph of $f^{(2)}$ for three values of k: slightly less than 3, equal to 3, and just greater than 3.

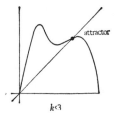 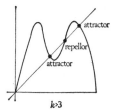

$k<3$	$k=3$	$k>3$

Figure 197a

We see that $f^{(2)}$ has a single fixed point for $k \leq 3$, and *three* fixed points for $k > 3$. Further, when $k \leq 3$ the fixed point is an attractor; for $k > 3$ the one in the middle becomes a repellor and the other two become attractors.

In other words, as k passes through 3 the original fixed point attractor splits into three pieces: a fixed point repellor and a periodic attractor of period 2. (Compare Figure 191 with $k = 3.05$. A spider's web diagram for this effect is shown in Figure 197b.)

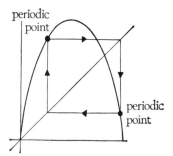

Figure 197b

This phenomenon is known as *bifurcation*, to capture the idea of the attractor 'splitting'. (One dynamical systems theorist insists the word should be 'furcation', since the splitting is not always – as here – into two bits.) We can illustrate this effect by drawing a new type of diagram: this graphs the values of x corresponding to periodic points against k, as follows:

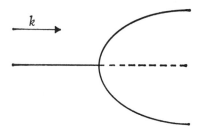

Figure 198

We shall call this a *bifurcation diagram*, to distinguish it from the graphs of f or its iterates. We show attractors as heavy lines, repellors as dotted lines (and later leave out the repellors to simplify the pictures).

As k increases further, the new periodic points also become repellors, by a similar process, but now occurring for $f^{(4)}$. The bifurcation diagram becomes:

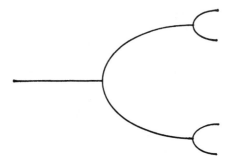

Figure 199

By the time we reach $k = 3\cdot57$ (approximately) this process has occurred infinitely often!

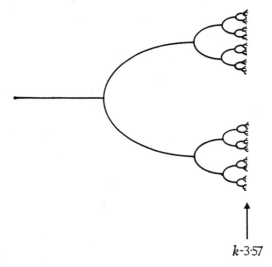

$k\sim3\cdot57$

Figure 200

What happens next? I will stop drawing explanatory pictures, since these start to get confusing and would take too much space; but Figure 200 makes it plausible that something unusual will occur. What we actually get, appearing simultaneously, are:

(a) An attractor of period *three*.

(b) Periodic repellors of *all* periods, all coexisting at the same time.

The result is that not all points settle down towards a steady behaviour: some of them lead to apparently random fluctuations as they are repelled from one repellor after another. This is the 'chaos' discovered by the ecologists[6] and hinted at in Figure 191 for $k \geq 3.7$. It is not quite as chaotic as one might suppose. For example, we can calculate[7] the numbers of periodic points of any given period (even though they are nearly all repellors). When $k = 3.832$, we have

　1 fixed point (other than 0),

　3 points of period dividing 2,

　4 points of period dividing 3,

　7 points of period dividing 4 (4 of period 4, 3 of period 2),

　11 points of period dividing 5,

　18 points of period dividing 6 (11 of period 6, 4 of period 3, 3 of period 2),

and so on, where the numbers in the left hand column are obtained by adding together the previous two, as for Fibonacci's series.[8] This result was obtained by the dynamical systems people, not by the ecologists!

However, up at $k = 4$ the chaos is more extreme. There are no periodic attractors at all. The whole unit interval becomes a *strange attractor*, and orbits jump around inside it in apparently random fashion (see Figure 191c). It doesn't just *look* random: it is possible to study the 'statistical mechanics' of the system and prove that it has exactly the statistical properties that one would find in a suitable random system.

What do I mean by 'random' here? It's a tricky question to answer, because the very phenomenon just discussed shows that it is not as simple an idea as might be imagined. 'Random' means that no obvious structure exists, but that 'on average' we can say various things, such as how often the values occur in a given range. But in the past, 'random' has carried the connotation of 'indeterminate', that is, a system is *deterministic* if it follows exactly some regular law, random if not.

However, here we have found a deterministic system (it obeys the formula $x_{n+1} = kx_n(1-x_n)$ *exactly*) whose behaviour is more sensibly described as random! In fact, if you were just presented with a list of iterates of some starting value and not *told* the formula, you would never guess that there *was* one. Thus the line between deterministic and random behaviour is no longer so clear-cut.

This carries implications for supposedly 'random' natural phenomena. Are they *really* random? For that matter, is this now a sensible question to ask? If not, what is?

Strange attractors and chaos do not just occur for this particular system. In a reasonable sense, they are the rule rather than the exception in dynamical systems; further, as the number of variables increases, the phenomena can become still more peculiar. There is a growing body of evidence that turbulence in fluid flow may be the manifestation of a strange attractor. Likewise, the fluctuations in the earth's magnetic field, including sudden switches of direction of magnetization, may be traced to the same source. The types of equation studied so assiduously by economists, who seem to be under the impression that these behave tamely, can actually exhibit chaotic effects of arbitrarily nasty kind. What the precise implications of all this will be for science, I cannot imagine; but it is quite clear that chaos, in this strictly technical sense, will increasingly make its presence felt.

Real mathematics

I'd like to end on a fairly serious note. In the present volume I have attempted to give some flavour of what mathematicians do and how they think about it. As far as possible I have tried not to water down the subject matter: I think it is reasonable to try to include quite advanced topics in a book of this nature, and I have found that readers generally seem to appreciate being given some inkling of what goes on in the higher realms of today's mathematics, even though this may occasion some difficulties of comprehension every so often. It has always seemed to me that the non-specialist is capable of grasping more of the spirit of a subject than is commonly imagined. Popular science magazines do not shy away from the abstruse complexity of elementary particle physics or molecular biology, and I can see no reason to shy away from comparable mathematical complexities.

Nonetheless, a popularization of a subject is not the real thing. The biggest danger with a popularization is that an insensitive reader may mistake the spirit for the substance. Topology, described as 'rubber-sheet geometry' and illustrated by a few striking examples, can reveal broad new vistas to a suitably attuned mind. On the other hand it can, if interpreted too literally and narrowly, look like the silliest nonsense imaginable. Stretching and bending indeed! What conceivable use can *that* be? Holes in doughnuts! Combing hairy balls! Two-dimensional creatures living on twisted bands! How utterly ridiculous.!

Popular books or articles always have to slide over what, to the professional, is perhaps the most important part of all of his subject; the hard, technical grind. Yes: the research worker usually works in a fashion which is far from logical, leaping to conclusions, leaving odd problems to be taken care of later, arguing by analogy, and making wild and unsubstantiated guesses. Yes: the rigorous, technical argument is usually filled in later. Yes: the intuitive ideas behind proofs are the true source of inspiration. Nonetheless, the technical grind is not just an afterthought: *it is an absolutely essential part of the process.* Without it, mathematics would collapse of its own weight. Without a firm framework of technique, and without the constraints imposed by rigour, the mathematician's intuition would lose its sharp edge, become dull and woolly, and fade into decadence.

Much scientific[9] methodology (and, if only in this technical sense, mathematics is far more akin to the sciences than it can ever be to the arts) has been developed to overcome fatal flaws in the human psyche, of which the worst is a willingness to believe something on too little evidence, simply because it would be nice if it were true. This is where mathematical rigour came from: particularly in the nineteenth century, attempts to dispense with it and argue by physical analogy or ill-founded principles derived from past habit came sadly to grief. Because scientific discoveries are so well founded, it has become possible to *build* on them. Newton remarked (and he was not being modest!) 'If I have seen further than others, it is because I have stood on the shoulders of giants'. In too many non-sciences, the aim seems more to be to stand on their feet.

The problem with the hard technical grind is that there is no way to appreciate it except by *doing* it. Unlike the general flavour and philosophy of the subject, it resists any attempt at paraphrase. All I can do, therefore, is ask you to recognize that there is another side to the subject from that presented above, and in particular that apparent frivolity may just be a way of rendering palatable the difficulties. Mathematicians and scientists habitually give complicated concepts simple names, or dress them up in whimsical imagery, as an aid to comprehension. This should not be confused with the use of jargon to obscure a basically simple idea – or one that evaporates completely if analysed too closely. 'Homotopy group' may sound like jargon but it is a relatively simple phrase for a complicated technical idea and much modern mathematics would not be possible without it.

The proper statement of the 'Hairy Ball Theorem' is that there is no continuous function $f:S^2 \to S^2$ which is homotopic to the identity but has no fixed point. Here S^2 is the two-dimensional sphere. To appreciate the technicalities of this statement involves a lot of preparatory work on continuity, topological spaces, fixed points, homotopy, and the like. To *prove* it involves a lot of algebraic topological technique.[10] Unless you have gone through all of this detail, it is not even clear that the Hairy Ball Theorem *says* anything mathematical at all. On the other hand, it is a tremendous help in grasping this mass of detail if you bear in mind that what the theorem says is that a hairy ball cannot be combed smooth! To do *mathematics* in this area, you need to understand *both* ways of thinking about the theorem *and* the connection between them *and* how to switch from one to the other to the greatest benefit.

Nor is this a lot of fuss over nothing. To the technician, the Hairy Ball Theorem is a key step in the development of methods whose practical implications include very powerful results on the existence of solutions to differential equations. You won't get *that* aspect just by thinking about combing the dog.

It is only my personal opinion, but I think that one of the worst offences committed by 'modern maths' in schools was to lose sight of this relation between the intuitive concepts and the technicalities. It may be very amusing to study topology at school, but unless this is very carefully motivated and taught by someone who understands where it fits in the Mathematical Scheme of Things, it will all seem vague and pointless; it will lack a satisfying solid core; it will be actually hard to comprehend *because* of this woolliness; and it may well put off able students, in the long run. But between the stereotypes of 'modern' and 'traditional' mathematics of school syllabuses lies the real thing, and the evidence for this is the unbroken development of genuine modern mathematics out of genuine traditional mathematics: they are both part of the same whole. There are already signs of increasing attention being paid to this middle ground and it is a tendency to be welcomed. If, at the same time, we can get away from the alarming habit that has grown up of shying away from anything that is the least bit difficult,[11] then we will make some *real* progress.

Notes

Chapter 1   Mathematics in General

1. I am told that in Dutch exactly the opposite usage prevails: the word now used in mathematics where we would use 'set' has for centuries been translated as 'collection'.

2. See W. Sierpinski, *On the Congruence of Sets and Their Equivalence by Finite Decomposition*, Lucknow University Studies, 1954; also E. Kasner and J. Newman, *Mathematics and the Imagination*, Bell, 1949.

3. To solve a polynomial equation

$$a_n x^n + \ldots + a_1 x + a_0 = 0$$

by radicals we must find a formula for the roots in terms of the coefficients a_0, a_1, a_n which uses only operations of addition, subtraction, multiplication, division, and extraction of roots. An example would be the standard solution of the quadratic equation

$$ax^2 + bx + c = 0$$

given by

$$x = \frac{-b \pm \sqrt{(b^2 - 4ac)}}{2a}.$$

It has been proved that no such expression exists for the roots of the general equation of degree 5. The proof is by means of Galois Theory and requires a good grounding in abstract algebra. For details see *Galois Theory*, E. Artin, Notre Dame, 1959; *Introduction to Field Theory*, I. T. Adamson, Oliver & Boyd, 1964; or *Galois Theory*, Ian Stewart, Chapman & Hall 1973.

Chapter 2   Motion without Movement

1. C. L. Dodgson, *Euclid and His Modern Rivals*, Macmillan, London, 1879, p. 48.

2. The rest may be found in F. Cajori, 'The History of Zeno's Arguments on Motion', *American Mathematical Monthly* 22, 1915; E. Kasner and J. Newman, *Mathematics and the Imagination*, Bell, 1949, and Russell, *Mysticism and Logic*.

3. This statement should perhaps be expanded. Given three points *A*, *B*, *C* (distinct and non-collinear) and three distances *a*, *b*, *c*, there is at most one point distance *a* from *A*, *b* from *B*, and *c* from *C*. For the circles of radius *a*, centre *A*, and radius *b*, centre *B* meet in at most two points: these are *not* equidistant from *C*, so at most one of them is distance *c* from *C*. Since rigid motions do not change distances, it follows that once we know what happens to a triangle we know what happens to everything else.

In three-dimensional space, we need an extra point, because instead of circles as above we have spheres.

4. One way round this is to write transformations on the right, as (*X*)*T* or *XT*. Then we can define

$$(X)EF = ((X)E)F$$

and everything comes out in order: *EF* means 'first *E*, then *F*'. This takes a little getting used to (though there are precedents, such as writing *n*! for factorials). Mathematicians often use this method.

Chapter 3 Short Cuts in the Higher Arithmetic

1. Although nowadays we first meet complex numbers in the solution of quadratic equations such as

$$x^2 + 3 = 0$$

where the usual formula gives $x = \pm\sqrt{-3}$, historically complex numbers first attracted attention in the solution of *cubic* equations.

Tartaglia's solution for the equation

$$x^3 + px + q = 0$$

is

$$x = \sqrt[3]{\left(\frac{-q}{2} + \sqrt{\left(\frac{p^3}{27} + \frac{q^2}{4}\right)}\right)} + \sqrt[3]{\left(\frac{-q}{2} - \sqrt{\left(\frac{p^3}{27} + \frac{q^2}{4}\right)}\right)} \quad .$$

For the equation

$$x^3 - 7x + 6 = 0 \tag{†}$$

we have $p = -7$, $q = 6$, which leads to

$$x = \sqrt[3]{\left(-3 + \frac{10i}{3\sqrt{3}}\right)} + \sqrt[3]{\left(-3 - \frac{10i}{3\sqrt{3}}\right)}$$

where $i = \sqrt{-1}$.

On the other hand, we can factorize (†) as

$$(x-1)(x-2)(x+3) = 0$$

so the solutions are $x = 1$, 2, or -3.

To reconcile the two results, observe that if we assume i behaves like a number we can work out the cube roots. For

$$\left(1+\frac{2i}{\sqrt{3}}\right)^3 = 1+\frac{3.2i}{\sqrt{3}}+\frac{3.4i^2}{3}+\frac{8i^3}{3\sqrt{3}}$$

$$= 1-4+\frac{18-8}{3\sqrt{3}}i$$

$$= -3+\frac{10i}{3\sqrt{3}},$$

and similarly

$$\left(1-\frac{2i}{\sqrt{3}}\right)^3 = -3-\frac{10i}{3\sqrt{3}}.$$

Using these for the cube roots we get

$$x = 1+\frac{2i}{\sqrt{3}}+1-\frac{2i}{\sqrt{3}}$$

$$= 2$$

which is one of the roots. Where are the other two? These come from different choices of cube roots: over the complex numbers, everything has *three* cube roots.

So by using complex numbers we end up with a correct result about *real* numbers. This is an impressive feat, which suggests that a study of complex numbers could be useful.

2. This phrase seems to be confined to 'modern maths' textbooks.

3. It's no problem. We are always free to make our *definitions* as we please, so long as we stick to them thereafter. In any case the notions of multiplication or subtraction of whole numbers, or of negative numbers, involve the same kind of method: forget where they came from, operate by analogy, and then check that the answers make sense.

4. Not the same 'congruent' as for triangles. But the two ideas have a certain amount in common: in each case one ignores differences of a specified kind: multiples of a fixed integer, or transformation by a rigid motion.

5. It is not hard to see that a number of the form 2^k+1 is not prime *unless* k is a power of 2. This is probably what led Fermat astray. Of all the unproved statements made by Fermat, this is the only one known to be false, and the only one about which he expressed doubt.

6. E. T. Bell, *Men of Mathematics*, vol. 1, Penguin Books, 1965, p. 73.

7. See Hardy and Wright, *An Introduction to the Theory of Numbers*, Oxford University Press, 1962, Chapter 6.

8. Fermat's 'last theorem' – so called because it is the last whose status is unresolved – asserts that there are no solutions of the equation

$$x^n + y^n = z^n$$

in non-zero integers x, y, z, provided that n is an integer greater than 2. For $n = 2$ there are an infinity of solutions, e.g. $3^2 + 4^2 = 5^2$.

[Addendum, 1994:] In June 1993 Andrew Wiles of Princeton University announced a proof of Fermat's Last Theorem, deriving it from a special case of the so-called Taniyama-Shimura conjecture. However, a technical flaw was subsequently found. At the start of 1994 the flaw was unrepaired, but toward the end of 1994 Wiles and Richard Taylor of Cambridge University announced they had completed the proof.

Chapter 4 The Language of Sets

1. The symbol is a stylized version of the Greek letter epsilon, the initial letter of the word 'element'. Many books (especially less recent ones) just use an epsilon.

2. A. A. Milne, 'Winnie-the-Pooh', Chapter 7.

3. This is Fermat's last theorem again. See note 8 to Chapter 3.

4. In some developments of arithmetic 0 is *defined* to be ∅, 1 to be {∅}, 2 to be {∅,{∅}}, and so on. So it is not completely true that 0 and ∅ are different. However, one certainly *thinks* about them in different ways, and this is what is important here.

5. The symbol '⊆' is derived from '≤', but is made curved as a reminder that it applies to sets, not to numbers. Some books use '⊂' instead.

6. The derivation of **N**, **R**, and **C** is clear. The **Z** comes from the German word *Zahl* (number). The **Q**, I think, is from 'quotient'. Sometimes **P** (for 'positive') is used instead of **N**.

7. Consider the set $C = \{\{1,2\},\{3,4\}\}$. Then $1 \in \{1,2\}$ and $\{1,2\} \in C$, but $1 \notin C$.

8. Sometimes referrred to as 'cup' and 'cap', which to my mind is confusing, and about as much help as calling '=' 'two parallel lines'. If you want a mnemonic, try '∪nion' and 'i∩tersection'.

9. The only place I have seen membership tables in print is in 'Venn Vill They Ever Learn?' by Frank Ellis, *Manifold* 6, 1970, p. 44.

10. Boolean algebra is of some use in the design of logic circuits for computers (see e.g. Rutherford, *Introduction to Lattice Theory*, Oliver & Boyd, 1965, pp. 31–40, 58–74). Apart from its connection with set theory it has little relevance to the main body of mathematics. However, quite a deep theory exists. See P. R. Halmos, *Lectures on Boolean Algebras*, Van Nostrand, 1963.

11. The game of 'vish' (short for 'vicious circle') is played by selecting at random a word in the dictionary, choosing a word from its

uennition, then choosing one from *its* definition, and so on – the object being to retrieve the starting word in as few steps as possible.

12. The first satisfactory definition of an ordered pair was given by Kuratowski in 1921. The difficulty is to avoid reference to the printed form of the symbols '(a, b)'. It will not do to say 'a is the left-hand element' because 'left' is not a concept of set theory. The early philosophers got themselves into a terrible tangle (see e.g. Russell, *The Principles of Mathematics*, 1903) over this question. 'Is the ordering a property of *a*?' No, it must depend on *b* as well, because the ordering of, say, (1,2) and (3,1) are different. 'Is it a property of *b*?' No, for the same reason. 'Is it a property of *a and b*?' No, because '*a* and *b*' is the same as '*b* and *a*', so that (*a,b*) and (*b,a*) come out the same.

The difficulty is to remove the symmetry between *a* and *b*. The philosophers were unable to do this because they were confused over the distinction between x and $\{x\}$. They wanted them to be the same. However, once we realize that they should be different, any number of paths are open. For instance, we can define

$$(a,b) = \{\{a\}, \{a,b\}\}.$$

The asymmetry on the right is enough to ensure that (*a,b*) and (*c,d*) are equal if and only if $a = c$ and $b = d$: this can be proved by elementary set theory. Since this is the only important property of ordered pairs, the definition is a suitable one. It is not one that is psychologically appealing, however.

Chapter 5   What is a Function?

1. This usage conflicts with current grammar, which would separate the word into 'on to'. It must therefore be thought of as a technical term in its own right if one is to preserve grammatical purity. I suspect it was coined in the U.S.A.

2. A purely set-theoretical definition of 'function' can be given, using the notion of 'ordered pair'. It usually causes students considerable difficulty, because it is at first sight far removed from the idea of a 'rule', and because they have met only numerical functions, defined by formulae.

The problem is to produce a single object f which ties each element x of the domain to the corresponding $f(x)$ in an unambiguous way. For any particular x, the ordered pair

$$(x, f(x))$$

does this rather well: for example if we have the pair (7,24) we know that $f(7) = 24$. To specify f over the whole domain D we can use the set whose elements are all the pairs $(x, f(x))$ as x runs through D.

If T is a target, all the elements $f(x)$ lie in T, so this set of ordered pairs is a subset of $D \times T$. By virtue of property (3) of functions (p. 67), this subset will satisfy two conditions:

(i) For every $x \in D$ there is some $y \in T$ such that (x,y) is in the given subset.

(ii) If (x,y) and (x,z) are in the subset, then $y = z$.

Condition (i) says that $f(x)$ is defined for all x, while (ii) ensures uniqueness.

Now we can turn the process on end. Given two sets D and T we define a *function from D to T* as a subset f of $D \times T$, subject to conditions (i) and (ii). If we are given $x \in D$ we find $f(x)$ by (a) finding some y with $(x,y) \in f$ (which we can do by (i)) and (b) defining

$$f(x) = y$$

which is unambiguous by (ii).

In other words, any 'rule' is tantamount to a set of ordered pairs. This is not immediately obvious (it would be tragic if everything in the world was) but, in the words of a well-known politician, 'You know it works.'

For further information consult Halmos, *Naive Set Theory*, Van Nostrand, 1964, or Hamilton and Landin, *Set Theory*, Prentice Hall, 1961.

Chapter 6 The Beginnings of Abstract Algebra

1. See note 3 for Chapter 1.

2. See Rouse Ball, *Mathematical Recreations and Essays*, Macmillan, 1959, Chapter 12.

3. Lindemann proved his theorem in 1882. See Stewart, *Galois Theory*, Chapman & Hall, 1973, p. 74.

4. Those readers who have not encountered complex numbers should consult W. W. Sawyer, *Mathematician's Delight*, Penguin Books, 1943.

5. See de Bruijn, 'A Solitaire Game and Its Relation to a Finite Field', *Journal of Recreational Mathematics* 5, 1972, p. 133.

Chapter 7 Symmetry: The Group Concept

1. See for example, Ledermann, *Introduction to Group Theory*, Oliver & Boyd, 1973.

There is a paper by J. H. Conway with the title 'A Group of Order 8 315 553 613 086 720 000', *Bulletin of the London Mathematical Society*

1, 1969, pp. 79–88. One could hardly define this group by a multiplication table! (It should not be thought that mathematicians spend their time constructing larger and larger groups. The size of Conway's group is not really important, but it has several rather remarkable properties which are.)

2. See Klein, *Lectures on the Icosahedron and the Solution of Equations of the Fifth Degree*, Kegan Paul, 1913 (also available in Dover Books).

3. See Coxeter, *Introduction to Geometry*, New York, Wiley, 1969, p. 278. This interesting book also has information about the 17 types of plane symmetry (pp. 50–61, 413).

Chapter 8   Axiomatics

1. In the real world, the points eventually get so close that one can no longer distinguish them. But we can also check the matter algebraically. For simplicity consider a radial *line*, and suppose that Γ has radius d. Any point strictly inside Γ is distance e from the centre, where e is *strictly less than d*. It follows that a point (say) distance $\frac{1}{2}(e+d)$ from the centre is still strictly inside Γ, but further out than distance e. For

$$d - \tfrac{1}{2}(e+d) = \tfrac{1}{2}(d-e) > 0,$$
$$\tfrac{1}{2}(e+d) - e = \tfrac{1}{2}(d-e) > 0.$$

2. W. W. Sawyer, *Prelude to Mathematics*, Penguin Books, 1955, p. 85.

Chapter 9   Counting: Finite and Infinite

1. Apart from certain set-theoretic difficulties, mentioned in Chapter 20.

2. Not without a lot of hard work! See Hamilton and Landin, *Set Theory*, Prentice Hall, 1961, pp. 133–238.

3. Birkhoff and MacLane, *A Survey of Modern Algebra*, Macmillan, 1963, p. 362.

Chapter 10   Topology

1. Assuming that all the materials have thickness, and that the objects are solid, the types are as follows: A, E, G, I are spheres; C,

D, *F* tori; *B*, *H* double tori. If (as in real life) *A*, *D*, *E* are hollow, and *I* is empty, there are more types: *A*, *E*, *I* are hollow spheres; *G* a solid sphere; *C*, *F* solid tori; *D* a hollow torus; *B*, *H* solid double tori. With more quibbles about details – bubbles in the bread, say – there are still more distinctions.

2. There is a science-fiction story by George Gamow ('The Heart on the Other Side', in *The Expert Dreamers*, edited by Frederik Pohl, Gollancz, 1963) in which it is hypothesized that the real universe is non-orientable. The hero is attempting to revolutionize the shoe-manufacturing industry, but runs into trouble when all his bodily proteins turn into mirror-image forms.

It is difficult to tell whether the physical universe is orientable. This is because orientability is what topologists call a 'global' property: to discover it one must look at the whole space. 'Locally' a Möbius band looks much like a cylinder: the properties close to any point are much the same. We have little information about the global structure of the universe since it is so large. But if the astronomical observations mentioned at the end of Chapter 8 are correct, it is probably *non-orientable*!

Chapter 11   The Power of Indirect Thinking

1. See Ore, *The Four-Colour Problem*, Academic Press, 1967.

Chapter 12   Topological Invariants

1. See E. C. Zeeman, *Introduction to Topology*, Penguin Books, forthcoming. I am grateful to Professor Zeeman for permission to incorporate his ideas.

2. See papers by Ringel and Youngs, *Proceedings of the National Academy of Science* (U.S.A.), 1968. Also A. T. White, *Graphs, Groups and Surfaces*, North-Holland/American Elsevier, Amsterdam, London and New York, 1973.

Chapter 13   Algebraic Topology

1. Another important class of algebraic invariants, the *homology groups*, first saw light in rudimentary numerical form, as the so-called *Betti numbers*.

Chapter 14   Into Hyperspace

1. See Rourke and Sanderson, *Piecewise Linear Topology*, Springer, 1972.

Chapter 15   Linear Algebra

1. There are any number of texts on linear algebra. There is a good, gentle introduction in W. W. Sawyer, *A Path to Modern Mathematics*, Penguin Books, 1966. From a practical viewpoint Fletcher, *Linear Algebra Through its Applications*, Van Nostrand, 1973, has much to recommend it.

2. W. W. Sawyer, *Prelude to Mathematics*, Penguin Books, 1955, Chapter 8.

Chapter 16   Real Analysis

1. The suspiciously vague-sounding phrases 'as small as we please' and 'sufficiently large' are in fact quite precise. The first means that given any positive number ε we can make $b_n - l$ smaller than ε. To do this we must make n larger than *some* number N, which may depend on ε. Thus a precise formulation of convergence runs: b_n tends to the limit l if for all $\varepsilon > 0$ there exists N such that if $n > N$

$$|b_n - l| < \varepsilon.$$

(The modulus signs ' | | ' just make sure that the difference is positive.)

2. The rearrangement of terms assumes a law which is not true even for convergent series – as the apparent paradox demonstrates! It can be shown that for convergent series with *positive* terms rearrangement is allowable.

3. As in note 1 the vagueness can be removed, and one gets the usual technical definition: the function f is continuous at the point x if for all $\varepsilon > 0$ there exists $\delta > 0$ such that

$$|f(x) - f(p)| < \varepsilon$$

whenever

$$|x - p| < \delta.$$

This may explain why students find analysis hard: the definition does not resemble one's intuitive idea that a continuous curve is one that is 'all in one piece'. The ε–δ definition does lend itself to proving that a given function, such as x^2, or $\sin(x)$, is continuous: this is why it is used. No one has yet found a simpler approach to the question without running into serious logical errors; but one should not imagine that the current theory of continuity is the last word on the subject.

Chapter 17 The Theory of Probability

1. See W. Weaver, *Lady Luck*, Doubleday, 1963 and D. Huff, *How to Take a Chance*, Penguin Books, 1977.

2. For example, consider the experiment of picking a random point on the real line. The probability of picking any particular number, such as 2, or π, is zero. But it is not *impossible* to pick 2, or π.

Chapter 18 Computers and Their Uses

1. There is a lot of interesting material in *Computers and Computation* (readings from *Scientific American*), Freeman, San Francisco.

2. For Fortran, try *The Elements of Fortran Style* by Kreitzberg and Schneiderman, Harcourt Brace Jovanovitch, New York, 1972, or *A Guide to Fortran Programming* by McCracken, Wiley, New York, 1961. For Algol see Wooldridge and Ratcliffe, *An Introduction to Algol Programming*, English Universities Press, 1963.

Chapter 19 Applications of Modern Mathematics

1. For example. Allen, *Mathematical Analysis for Economists*, Macmillan, 1970 and A. Battersby, *Mathematics in Management*, Penguin Books, 1966.

2. For more details see the article 'Mathematics in the Physical Sciences' by F. J. Dyson, in *Mathematics in the Modern World* (readings from *Scientific American*) edited by Morris Kline, published by Freeman, San Francisco, 1970.

3. See R. Thom, translated by D. Fowler, *Structural Stability & Morphogensis*, Benjamin, Reading, Massachusetts, 1975. Since I wrote this chapter several accessible references for Thom's theory have appeared. Among them are:

E. C. Zeeman, 'Catastrophe Theory', *Scientific American* 234, 1976, pp. 65–83.

I. N. Stewart, 'The Seven Elementary Catastrophes', *New Scientist* 68, 1975 pp., 447–54.

T. Poston and I. N. Stewart, 'Taylor Expansions and Catastrophes', *Research Notes in Mathematics* 7, Pitman Publishing, London, 1976.

T. Poston and I. N. Stewart, *Catastrophe Theory and its Applications*, Pitman Publishing, London, 1978.

E. C. Zeeman, *Catastrophe Theory: Selected Papers (1972–1977)*, Addison-Wesley, Reading, Massachusetts, 1977.

4. See Poston and Woodcock, 'Zeeman's Catastrophe Machine', *Proceedings of the Cambridge Philosophical Society* 74, 1973, pp. 217–26.

5. A large number of computer drawings of catastrophes appear in 'A Geometrical Study of the Elementary Catastrophes', Woodcock and Poston, *Lecture Notes in Mathematics* 373, Springer, 1974.

Chapter 20 Foundations

1. See A. Phillips, 'Turning a Surface Inside Out', *Scientific American*, May 1966, pp. 112–20.

2. No names, no pack-drill!

3. . . . . who shaves everyone who does not shave himself. Who shaves the barber?

4. Namely, Von Neumann–Bernays–Gödel set theory. See Bernays and Fraenkel, *Axiomatic Set Theory*, North Holland, 1958, p. 31.

5. 'On Formally Undecidable Propositions of *Principia Mathematica* and Related Systems I', *Monatshefte für Mathematik und Physik* 38, 1931, pp. 173–98. See note 6.

6. Gödel's paper, together with a commentary, is available in translation under the original title (Meltzer and Braithwaite, published by Oliver & Boyd, 1962).

7. See Matijasevič, 'Enumerable Sets are Diophantine', *Soviet Mathematics* [*Doklady*] 11, 1970, pp. 354–8; 'Diophantine Representation of the Set oɪ Prime Numbers', ibid. 12, 1971, pp. 249–54.

8. Much effort has been expended on a search for formulae which represent all primes, or, at least, take only prime values. (No polynomial can represent the prime numbers and nothing else.) Thus we have Fermat's attempt with

$$2^{2^{n}}+1$$

or Euler's

$$n^2-79n+1601$$

which is prime for $n = 0, \ldots, 79$ (but composite for $n = 80$). There are various ways of cheating by choosing an interpretation of the word 'formula'. Many references are given by Dudley, 'History of a Formula for Primes', *American Mathematical Monthly* 76, 1969, p. 23.

Such formulae are unlikely to be of use in the study of prime numbers because they yield little real insight: a formula is usually more intractable than a simple non-formal definition. Matijasevič's results should be thought of as evidence for the complexity of polynomials, rather than simplicity of prime numbers.

See, in general, G. H. Hardy, *Pure Mathematics*, Cambridge University Press, 1959 or H. Davenport, *Higher Arithmetic*, Humanities, 1968.

9. In fact, Gödel proved that assuming the truth of the Continuum hypothesis will not lead to contradictions in set theory (*The Consistency of the Continuum Hypothesis*, Princeton, New Jersey, 1940). Cohen proved that assuming its falsity likewise introduced no contradictions (*Set Theory and the Continuum Hypothesis*, Benjamin).

10. See Mendelson, *Introduction to Mathematical Logic*, Van Nostrand, 1964.

Appendix   And still it moves . . .

1. See K. Appel and W. Haken, 'Every Planar Map is Four Colorable', *Bulletin of the American Mathematical Society* 82, 1976, pp. 711–12. The full proof is to appear in the *Illinois Journal of Mathematics*.

2. See J. P. Jones, D. Sato, H. Wada, and D. Wiens, 'Diophantine Representation of the Set of Prime Numbers', *American Mathematical Monthly* 83, 1976, pp. 449–64.

3. Namely M. Davis, Y. Matijasevič, and J. Robinson, 'Hilbert's Tenth Problem. Diophantine Equations: Positive Aspects of a Negative Solution', in *Proceedings of Symposia in Pure Mathematics* 28, *Mathematical Developments Arising from Hilbert Problems*, American Mathematical Society, 1976, pp. 323–78.

4. According to M. V. Berry, *Principles of Cosmology and Gravitation*, Cambridge University Press, 1976, p. 122, the known observable universe has diameter roughly 10^{26} m, hence volume 4×10^{78} m^3. The volume of a rabbit is at least 10^{-3} m^3. The total volume of rabbits equals that of the universe when

$$100 \times 5^T \times 10^{-3} = 4 \times 10^{78},$$

or

$$5^T = 4 \times 10^{79}$$
$$T \log 5 = 79 + \log 4 \quad \text{(logs base 10)}$$
$$T = 114.$$

After at most 114 generations, we have more rabbit than universe.

5. See J. Guckenheimer, G. Oster, and A. Ipaktchi, '*The Dynamics of Density-dependent Population Models*', to appear.

6. See T.-Y. Li and J. A. Yorke, 'Period Three Implies Chaos', *American Mathematical Monthly* 82, 1975, pp. 985–92.

7. See S. Smale and R. F. Williams, 'The Qualitative Analysis of a Difference Equation of Population Growth', *Journal of Mathematical Biology* 3, 1976, pp. 1–4.

8. The sequence 1,1,2,3,5,8,13,21, . . . , in which each number after the first two is the sum of the previous two, was introduced by Leonardo of Pisa ('Fibonacci') in *c.* 1225, in a problem about the progeny of rabbits. (The recurrence relation is now $x_{n+1} = x_n + x_{n+1}$; the reader may reconstruct the problem from this.)

9. I exclude a number of self-styled 'sciences' of recent vintage, whose assumption of that term has greatly devalued it.

10. See for example C. R. F. Maunder, *Algebraic Topology*, Van Nostrand, London, 1970, p. 131.

11. Such as the proposal to reverse the trend (current in the mid 1970s but abating at the time of writing) away from science: *make science easier*! This 'solution' to the problem was based on the belief – probably true – that students studied science less because of the wide variety of newer and easier subjects available. The fallacy should be obvious: make science easier and the 'scientists' you produce don't know enough to be worthy of that name, hence are useless for all practical purposes. (Only someone who *failed* to appreciate my point about the need for 'hard, technical grind' could imagine that it is possible to dilute the content of a science course without severe ill-effects.) There is an alternative solution that springs to mind, however: I won't suggest it explicitly, but I urge the reader to consider Jacobi's dictum: '*always invert*'.

Glossary of Symbols

$\equiv$	congruence (of integers)
$(\bmod\ c)$	to the modulus c
$\in$	membership
$\{\ \}$	the set whose members are ...
$\{x \mid \}$	the set of all x such that ...
$\emptyset$	the empty set
$\subseteq$	inclusion of sets
N	the set of natural numbers
Z	the set of integers
Q	the set of rational numbers
R	the set of real numbers
C	the set of complex numbers
$\cup$	union
$\cap$	intersection
$-$	difference (of sets)
V	the universal set
S'	set-theoretic complement
$A \times B$	Cartesian product of sets
(a,b)	ordered pair
$\mathbf{R}^2$	the Euclidean plane
$x!$	factorial x $(= x.\,(x-1)\,(x-2) \ldots 3.2.1)$
$f\!:\!D \to T$	a function f from D to T
fg	multiplication of functions
1_D	the identity function on D
K	the set of constructible numbers
π	$3 \cdot 14159 \ldots$
$[x]$	the set of integers congruent to x
$\mathbf{R}[x]$	the set of polynomials in x
Σ	summation sign
$*$	group operation
I	identity element of a group
x'	inverse element to x
x^{-1}	inverse element to x
$\aleph_0, \aleph_1, \aleph_2, \ldots$	infinite cardinals
$a \leq \beta$	inequality of cardinals

$a < \beta$	strict inequality of cardinals
c	the cardinal of the set of real numbers
e	$2 \cdot 71828 \ldots$
$\lvert x \rvert$	the absolute value of x ($= x$ if $x \geq 0$, $-x$ if $x < 0$)
F	the number of faces in a map
V	the number of vertices in a map
E	the number of edges in a map
$\chi(S)$	the Euler characteristic of a surface
$\chi(N)$	the Euler characteristic of a network
$[\ ]$	the greatest integer not greater than ...
$p^{*}q$	composition of paths
$[p]$	the homotopy class of p
$\pi(S)$	the fundamental group of S
$\mathbf{R}^{3}, \mathbf{R}^{4}, \mathbf{R}^{5}, \mathbf{R}^{n}$	space of $3,4,5,n$ dimensions
$\pi_{n}(S)$	the nth homotopy group of S
$\begin{pmatrix} a & b \\ c & d \end{pmatrix}$	matrix
$\begin{pmatrix} x \\ y \end{pmatrix}$	column vector
$p(E)$	probability of event E
$\begin{pmatrix} n \\ r \end{pmatrix}$	binomial coefficient
B	the Russell set
$\mathscr{S}$	an axiomatic set theory
$\mathscr{A}$	ordinary arithmetic
$[a{:}t]$	the result of substituting t in a
$R(n)$	the nth sign enumerated by Gödel numbers

Index

About the Author

IAN STEWART was born in 1945. He was an undergraduate at Cambridge University and a PhD student at the University of Warwick, where he is now a Professor of Mathematics. He has held visiting positions in Germany (Tübingen), New Zealand (Auckland), and the USA (Storrs, CT and Houston, TX).

Dr. Stewart is an active research mathematician with over 90 published papers. His present interests centre upon the effects of symmetry on dynamics, with applications to pattern formation and chaos, in areas including fluid dynamics, mathematical biology, chemical reactions, electronic circuits, and animal locomotion. He takes a particular interest in problems that lie in the gaps between pure and applied mathematics. He is Director of Warwick University's Interdisciplinary Mathematical Research Programme and coordinator of the European Bifurcation Theory Group, a research network. He is the author of several research texts including *Singularities and Groups in Bifurcation Theory* (with Martin Golubitsky and David Schaeffer) and *Catastrophe Theory and Its Applications* (with Tim Poston).

His current writing is centred on popular science. His most recent book is *The Collapse of Chaos*, written jointly with the biologist Jack Cohen, which is a mix of mathematics, developmental biology, evolutionary theory, and anthropology. He has published over 60 books, including *Does God Play Dice?* (which has sold over 100,000 copies in English and has been translated into thirteen other languages), *The Problems of Mathematics, Game, Set & Math, Another Fine Math You've Got Me Into,* and (with Martin Golubitsky) *Fearful Symmetry: Is God a Geometer?* Earlier books include three mathematical comic books published in French: *Oh! Catastrophe!, Les Fractals,* and *Ah! Les Beaux Groupes.*

He contributes to a wide range of newspapers and magazines in Europe and the USA, including *New Scientist, Scientific American, The Sciences,* and *Discover.* He is mathematics consultant for *New Scientist* and a consultant for the *Encyclopædia Britannica.* He writes the bimonthly 'Mathematical Recreations' column in *Scientific American,* plus extra columns for the French, German, and Spanish editions. He also writes science fiction, with 18 short stories published in *Omni* and *Analog.*

He appears frequently on British radio. He scripted and presented a British Broadcasting Corporation radio program *Chaos!,* and he presented a regular puzzle slot on the Canadian Broadcasting Corporation's *Quirks and Quarks.* He has appeared on British TV in the *Equinox* (shown in the US as *Nova*) programmes 'Chaos' and 'Antichaos', *The Late Show,* an *Antenna* programme on the 'two cultures', a science series *Reality on the Rocks,* a debate on *Women in Science,* and two independently produced programmes *The Colours of Infinity* and *Chaos.*

He is a member of the American Mathematical Society, Science Fiction and Fantasy Writers of America, American Association for the Advancement of Science, Mathematical Association of America, American Museum of Natural History, New York Academy of Sciences, International Society for the Interdisciplinary Study of Symmetry, Cambridge Philosophical Society, London Mathematical Society, and the Institute of Mathematics and its Applications.

A CATALOG OF SELECTED
DOVER BOOKS
IN ALL FIELDS OF INTEREST

A CATALOG OF SELECTED DOVER
BOOKS IN ALL FIELDS OF INTEREST

CONCERNING THE SPIRITUAL IN ART, Wassily Kandinsky. Pioneering work by father of abstract art. Thoughts on color theory, nature of art. Analysis of earlier masters. 12 illustrations. 80pp. of text. 5⅜ × 8½.　23411-8 Pa. $3.95

ANIMALS: 1,419 Copyright-Free Illustrations of Mammals, Birds, Fish, Insects, etc., Jim Harter (ed.). Clear wood engravings present, in extremely lifelike poses, over 1,000 species of animals. One of the most extensive pictorial sourcebooks of its kind. Captions. Index. 284pp. 9 × 12.　23766-4 Pa. $12.95

CELTIC ART: The Methods of Construction, George Bain. Simple geometric techniques for making Celtic interlacements, spirals, Kells-type initials, animals, humans, etc. Over 500 illustrations. 160pp. 9 × 12. (USO)　22923-8 Pa. $9.95

AN ATLAS OF ANATOMY FOR ARTISTS, Fritz Schider. Most thorough reference work on art anatomy in the world. Hundreds of illustrations, including selections from works by Vesalius, Leonardo, Goya, Ingres, Michelangelo, others. 593 illustrations. 192pp. 7⅛ × 10¼.　20241-0 Pa. $9.95

CELTIC HAND STROKE-BY-STROKE (Irish Half-Uncial from "The Book of Kells"): An Arthur Baker Calligraphy Manual, Arthur Baker. Complete guide to creating each letter of the alphabet in distinctive Celtic manner. Covers hand position, strokes, pens, inks, paper, more. Illustrated. 48pp. 8¼ × 11.
24336-2 Pa. $3.95

EASY ORIGAMI, John Montroll. Charming collection of 32 projects (hat, cup, pelican, piano, swan, many more) specially designed for the novice origami hobbyist. Clearly illustrated easy-to-follow instructions insure that even beginning papercrafters will achieve successful results. 48pp. 8¼ × 11.　27298-2 Pa. $2.95

THE COMPLETE BOOK OF BIRDHOUSE CONSTRUCTION FOR WOOD-WORKERS, Scott D. Campbell. Detailed instructions, illustrations, tables. Also data on bird habitat and instinct patterns. Bibliography. 3 tables. 63 illustrations in 15 figures. 48pp. 5¼ × 8½.　24407-5 Pa. $1.95

BLOOMINGDALE'S ILLUSTRATED 1886 CATALOG: Fashions, Dry Goods and Housewares, Bloomingdale Brothers. Famed merchants' extremely rare catalog depicting about 1,700 products: clothing, housewares, firearms, dry goods, jewelry, more. Invaluable for dating, identifying vintage items. Also, copyright-free graphics for artists, designers. Co-published with Henry Ford Museum & Green-field Village. 160pp. 8¼ × 11.　25780-0 Pa. $9.95

HISTORIC COSTUME IN PICTURES, Braun & Schneider. Over 1,450 costumed figures in clearly detailed engravings—from dawn of civilization to end of 19th century. Captions. Many folk costumes. 256pp. 8⅜ × 11¾.　23150-X Pa. $11.95

STICKLEY CRAFTSMAN FURNITURE CATALOGS, Gustav Stickley and L. & J. G. Stickley. Beautiful, functional furniture in two authentic catalogs from 1910. 594 illustrations, including 277 photos, show settles, rockers, armchairs, reclining chairs, bookcases, desks, tables. 183pp. 6½ × 9¼. 23838-5 Pa. $9.95

AMERICAN LOCOMOTIVES IN HISTORIC PHOTOGRAPHS: 1858 to 1949, Ron Ziel (ed.). A rare collection of 126 meticulously detailed official photographs, called "builder portraits," of American locomotives that majestically chronicle the rise of steam locomotive power in America. Introduction. Detailed captions. xi + 129pp. 9 × 12. 27393-8 Pa. $12.95

AMERICA'S LIGHTHOUSES: An Illustrated History, Francis Ross Holland, Jr. Delightfully written, profusely illustrated fact-filled survey of over 200 American lighthouses since 1716. History, anecdotes, technological advances, more. 240pp. 8 × 10¾. 25576-X Pa. $11.95

TOWARDS A NEW ARCHITECTURE, Le Corbusier. Pioneering manifesto by founder of "International School." Technical and aesthetic theories, views of industry, economics, relation of form to function, "mass-production split" and much more. Profusely illustrated. 320pp. 6⅛ × 9¼. (USO) 25023-7 Pa. $9.95

HOW THE OTHER HALF LIVES, Jacob Riis. Famous journalistic record, exposing poverty and degradation of New York slums around 1900, by major social reformer. 100 striking and influential photographs. 233pp. 10 × 7⅞. 22012-5 Pa $10.95

FRUIT KEY AND TWIG KEY TO TREES AND SHRUBS, William M. Harlow. One of the handiest and most widely used identification aids. Fruit key covers 120 deciduous and evergreen species; twig key 160 deciduous species. Easily used. Over 300 photographs. 126pp. 5⅜ × 8½. 20511-8 Pa. $3.95

COMMON BIRD SONGS, Dr. Donald J. Borror. Songs of 60 most common U.S. birds: robins, sparrows, cardinals, bluejays, finches, more—arranged in order of increasing complexity. Up to 9 variations of songs of each species. Cassette and manual 99911-4 $8.95

ORCHIDS AS HOUSE PLANTS, Rebecca Tyson Northen. Grow cattleyas and many other kinds of orchids—in a window, in a case, or under artificial light. 63 illustrations. 148pp. 5⅜ × 8½. 23261-1 Pa. $4.95

MONSTER MAZES, Dave Phillips. Masterful mazes at four levels of difficulty. Avoid deadly perils and evil creatures to find magical treasures. Solutions for all 32 exciting illustrated puzzles. 48pp. 8¼ × 11. 26005-4 Pa. $2.95

MOZART'S DON GIOVANNI (DOVER OPERA LIBRETTO SERIES), Wolfgang Amadeus Mozart. Introduced and translated by Ellen H. Bleiler. Standard Italian libretto, with complete English translation. Convenient and thoroughly portable—an ideal companion for reading along with a recording or the performance itself. Introduction. List of characters. Plot summary. 121pp. 5¼ × 8½. 24944-1 Pa. $2.95

TECHNICAL MANUAL AND DICTIONARY OF CLASSICAL BALLET, Gail Grant. Defines, explains, comments on steps, movements, poses and concepts. 15-page pictorial section. Basic book for student, viewer. 127pp. 5⅜ × 8½. 21843-0 Pa. $4.95

BRASS INSTRUMENTS: Their History and Development, Anthony Baines. Authoritative, updated survey of the evolution of trumpets, trombones, bugles, cornets, French horns, tubas and other brass wind instruments. Over 140 illustrations and 48 music examples. Corrected and updated by author. New preface. Bibliography. 320pp. 5⅜ × 8½. 27574-4 Pa. $9.95

HOLLYWOOD GLAMOR PORTRAITS, John Kobal (ed.). 145 photos from 1926–49. Harlow, Gable, Bogart, Bacall; 94 stars in all. Full background on photographers, technical aspects. 160pp. 8⅜ × 11¼. 23352-9 Pa. $11.95

MAX AND MORITZ, Wilhelm Busch. Great humor classic in both German and English. Also 10 other works: "Cat and Mouse," "Plisch and Plumm," etc. 216pp. 5⅜ × 8½. 20181-3 Pa. $5.95

THE RAVEN AND OTHER FAVORITE POEMS, Edgar Allan Poe. Over 40 of the author's most memorable poems: "The Bells," "Ulalume," "Israfel," "To Helen," "The Conqueror Worm," "Eldorado," "Annabel Lee," many more. Alphabetic lists of titles and first lines. 64pp. 5³⁄₁₆ × 8¼. 26685-0 Pa. $1.00

SEVEN SCIENCE FICTION NOVELS, H. G. Wells. The standard collection of the great novels. Complete, unabridged. First Men in the Moon, Island of Dr. Moreau, War of the Worlds, Food of the Gods, Invisible Man, Time Machine, In the Days of the Comet. Total of 1,015pp. 5⅜ × 8½. (USO) 20264-X Clothbd. $29.95

AMULETS AND SUPERSTITIONS, E. A. Wallis Budge. Comprehensive discourse on origin, powers of amulets in many ancient cultures: Arab, Persian, Babylonian, Assyrian, Egyptian, Gnostic, Hebrew, Phoenician, Syriac, etc. Covers cross, swastika, crucifix, seals, rings, stones, etc. 584pp. 5⅜ × 8½. 23573-4 Pa. $12.95

RUSSIAN STORIES/PYCCKNE PACCKA3bl: A Dual-Language Book, edited by Gleb Struve. Twelve tales by such masters as Chekhov, Tolstoy, Dostoevsky, Pushkin, others. Excellent word-for-word English translations on facing pages, plus teaching and study aids, Russian/English vocabulary, biographical/critical introductions, more. 416pp. 5⅜ × 8½. 26244-8 Pa. $8.95

PHILADELPHIA THEN AND NOW: 60 Sites Photographed in the Past and Present, Kenneth Finkel and Susan Oyama. Rare photographs of City Hall, Logan Square, Independence Hall, Betsy Ross House, other landmarks juxtaposed with contemporary views. Captures changing face of historic city. Introduction. Captions. 128pp. 8¼ × 11. 25790-8 Pa. $9.95

AIA ARCHITECTURAL GUIDE TO NASSAU AND SUFFOLK COUNTIES, LONG ISLAND, The American Institute of Architects, Long Island Chapter, and the Society for the Preservation of Long Island Antiquities. Comprehensive, well-researched and generously illustrated volume brings to life over three centuries of Long Island's great architectural heritage. More than 240 photographs with authoritative, extensively detailed captions. 176pp. 8¼ × 11. 26946-9 Pa. $14.95

NORTH AMERICAN INDIAN LIFE: Customs and Traditions of 23 Tribes, Elsie Clews Parsons (ed.). 27 fictionalized essays by noted anthropologists examine religion, customs, government, additional facets of life among the Winnebago, Crow, Zuni, Eskimo, other tribes. 480pp. 6⅛ × 9¼. 27377-6 Pa. $10.95

FRANK LLOYD WRIGHT'S HOLLYHOCK HOUSE, Donald Hoffmann. Lavishly illustrated, carefully documented study of one of Wright's most controversial residential designs. Over 120 photographs, floor plans, elevations, etc. Detailed perceptive text by noted Wright scholar. Index. 128pp. 9¼ × 10¾.

27133-1 Pa. $11.95

THE MALE AND FEMALE FIGURE IN MOTION: 60 Classic Photographic Sequences, Eadweard Muybridge. 60 true-action photographs of men and women walking, running, climbing, bending, turning, etc., reproduced from rare 19th-century masterpiece. vi + 121pp. 9 × 12. 24745-7 Pa. $10.95

1001 QUESTIONS ANSWERED ABOUT THE SEASHORE, N. J. Berrill and Jacquelyn Berrill. Queries answered about dolphins, sea snails, sponges, starfish, fishes, shore birds, many others. Covers appearance, breeding, growth, feeding, much more. 305pp. 5¼ × 8¼. 23366-9 Pa. $7.95

GUIDE TO OWL WATCHING IN NORTH AMERICA, Donald S. Heintzelman. Superb guide offers complete data and descriptions of 19 species: barn owl, screech owl, snowy owl, many more. Expert coverage of owl-watching equipment, conservation, migrations and invasions, etc. Guide to observing sites. 84 illustrations. xiii + 193pp. 5⅜ × 8½. 27344-X Pa. $8.95

MEDICINAL AND OTHER USES OF NORTH AMERICAN PLANTS: A Historical Survey with Special Reference to the Eastern Indian Tribes, Charlotte Erichsen-Brown. Chronological historical citations document 500 years of usage of plants, trees, shrubs native to eastern Canada, northeastern U.S. Also complete identifying information. 343 illustrations. 544pp. 6½ × 9¼. 25951-X Pa. $12.95

STORYBOOK MAZES, Dave Phillips. 23 stories and mazes on two-page spreads: Wizard of Oz, Treasure Island, Robin Hood, etc. Solutions. 64pp. 8¼ × 11.

23628-5 Pa. $2.95

NEGRO FOLK MUSIC, U.S.A., Harold Courlander. Noted folklorist's scholarly yet readable analysis of rich and varied musical tradition. Includes authentic versions of over 40 folk songs. Valuable bibliography and discography. xi + 324pp. 5⅜ × 8½. 27350-4 Pa. $7.95

MOVIE-STAR PORTRAITS OF THE FORTIES, John Kobal (ed.). 163 glamor, studio photos of 106 stars of the 1940s: Rita Hayworth, Ava Gardner, Marlon Brando, Clark Gable, many more. 176pp. 8⅜ × 11¼. 23546-7 Pa. $11.95

BENCHLEY LOST AND FOUND, Robert Benchley. Finest humor from early 30s, about pet peeves, child psychologists, post office and others. Mostly unavailable elsewhere. 73 illustrations by Peter Arno and others. 183pp. 5⅜ × 8½.

22410-4 Pa. $5.95

YEKL and THE IMPORTED BRIDEGROOM AND OTHER STORIES OF YIDDISH NEW YORK, Abraham Cahan. Film Hester Street based on Yekl (1896). Novel, other stories among first about Jewish immigrants on N.Y.'s East Side. 240pp. 5⅜ × 8½. 22427-9 Pa. $6.95

SELECTED POEMS, Walt Whitman. Generous sampling from *Leaves of Grass.* Twenty-four poems include "I Hear America Singing," "Song of the Open Road," "I Sing the Body Electric," "When Lilacs Last in the Dooryard Bloom'd," "O Captain! My Captain!"—all reprinted from an authoritative edition. Lists of titles and first lines. 128pp. 5³⁄₁₆ × 8¼. 26878-0 Pa. $1.00

THE BEST TALES OF HOFFMANN, E. T. A. Hoffmann. 10 of Hoffmann's most important stories: "Nutcracker and the King of Mice," "The Golden Flowerpot," etc. 458pp. 5⅜ × 8½. 21793-0 Pa. $8.95

FROM FETISH TO GOD IN ANCIENT EGYPT, E. A. Wallis Budge. Rich detailed survey of Egyptian conception of "God" and gods, magic, cult of animals, Osiris, more. Also, superb English translations of hymns and legends. 240 illustrations. 545pp. 5⅜ × 8½. 25803-3 Pa. $11.95

FRENCH STORIES/CONTES FRANÇAIS: A Dual-Language Book, Wallace Fowlie. Ten stories by French masters, Voltaire to Camus: "Micromegas" by Voltaire; "The Atheist's Mass" by Balzac; "Minuet" by de Maupassant; "The Guest" by Camus, six more. Excellent English translations on facing pages. Also French-English vocabulary list, exercises, more. 352pp. 5⅜ × 8½. 26443-2 Pa. $8.95

CHICAGO AT THE TURN OF THE CENTURY IN PHOTOGRAPHS: 122 Historic Views from the Collections of the Chicago Historical Society, Larry A. Viskochil. Rare large-format prints offer detailed views of City Hall, State Street, the Loop, Hull House, Union Station, many other landmarks, circa 1904–1913. Introduction. Captions. Maps. 144pp. 9⅜ × 12¼. 24656-6 Pa. $12.95

OLD BROOKLYN IN EARLY PHOTOGRAPHS, 1865–1929, William Lee Younger. Luna Park, Gravesend race track, construction of Grand Army Plaza, moving of Hotel Brighton, etc. 157 previously unpublished photographs. 165pp. 8⅞ × 11¾. 23587-4 Pa. $13.95

THE MYTHS OF THE NORTH AMERICAN INDIANS, Lewis Spence. Rich anthology of the myths and legends of the Algonquins, Iroquois, Pawnees and Sioux, prefaced by an extensive historical and ethnological commentary. 36 illustrations. 480pp. 5⅜ × 8½. 25967-6 Pa. $8.95

AN ENCYCLOPEDIA OF BATTLES: Accounts of Over 1,560 Battles from 1479 B.C. to the Present, David Eggenberger. Essential details of every major battle in recorded history from the first battle of Megiddo in 1479 B.C. to Grenada in 1984. List of Battle Maps. New Appendix covering the years 1967–1984. Index. 99 illustrations. 544pp. 6½ × 9¼. 24913-1 Pa. $14.95

SAILING ALONE AROUND THE WORLD, Captain Joshua Slocum. First man to sail around the world, alone, in small boat. One of great feats of seamanship told in delightful manner. 67 illustrations. 294pp. 5⅜ × 8½. 20326-3 Pa. $5.95

ANARCHISM AND OTHER ESSAYS, Emma Goldman. Powerful, penetrating, prophetic essays on direct action, role of minorities, prison reform, puritan hypocrisy, violence, etc. 271pp. 5⅜ × 8½. 22484-8 Pa. $5.95

MYTHS OF THE HINDUS AND BUDDHISTS, Ananda K. Coomaraswamy and Sister Nivedita. Great stories of the epics; deeds of Krishna, Shiva, taken from puranas, Vedas, folk tales; etc. 32 illustrations. 400pp. 5⅜ × 8½. 21759-0 Pa. $9.95

BEYOND PSYCHOLOGY, Otto Rank. Fear of death, desire of immortality, nature of sexuality, social organization, creativity, according to Rankian system. 291pp. 5⅜ × 8½. 20485-5 Pa. $8.95

A THEOLOGICO-POLITICAL TREATISE, Benedict Spinoza. Also contains unfinished Political Treatise. Great classic on religious liberty, theory of government on common consent. R. Elwes translation. Total of 421pp. 5⅜ × 8½.
20249-6 Pa. $8.95

MY BONDAGE AND MY FREEDOM, Frederick Douglass. Born a slave, Douglass became outspoken force in antislavery movement. The best of Douglass' autobiographies. Graphic description of slave life. 464pp. 5⅜ × 8½. 22457-0 Pa. $8.95

FOLLOWING THE EQUATOR: A Journey Around the World, Mark Twain. Fascinating humorous account of 1897 voyage to Hawaii, Australia, India, New Zealand, etc. Ironic, bemused reports on peoples, customs, climate, flora and fauna, politics, much more. 197 illustrations. 720pp. 5⅜ × 8½. 26113-1 Pa. $15.95

THE PEOPLE CALLED SHAKERS, Edward D. Andrews. Definitive study of Shakers: origins, beliefs, practices, dances, social organization, furniture and crafts, etc. 33 illustrations. 351pp. 5⅜ × 8½. 21081-2 Pa. $8.95

THE MYTHS OF GREECE AND ROME, H. A. Guerber. A classic of mythology, generously illustrated, long prized for its simple, graphic, accurate retelling of the principal myths of Greece and Rome, and for its commentary on their origins and significance. With 64 illustrations by Michelangelo, Raphael, Titian, Rubens, Canova, Bernini and others. 480pp. 5⅜ × 8½. 27584-1 Pa. $9.95

PSYCHOLOGY OF MUSIC, Carl E. Seashore. Classic work discusses music as a medium from psychological viewpoint. Clear treatment of physical acoustics, auditory apparatus, sound perception, development of musical skills, nature of musical feeling, host of other topics. 88 figures. 408pp. 5⅜ × 8½. 21851-1 Pa. $9.95

THE PHILOSOPHY OF HISTORY, Georg W. Hegel. Great classic of Western thought develops concept that history is not chance but rational process, the evolution of freedom. 457pp. 5⅜ × 8½. 20112-0 Pa. $9.95

THE BOOK OF TEA, Kakuzo Okakura. Minor classic of the Orient: entertaining, charming explanation, interpretation of traditional Japanese culture in terms of tea ceremony. 94pp. 5⅜ × 8½. 20070-1 Pa. $3.95

LIFE IN ANCIENT EGYPT, Adolf Erman. Fullest, most thorough, detailed older account with much not in more recent books, domestic life, religion, magic, medicine, commerce, much more. Many illustrations reproduce tomb paintings, carvings, hieroglyphs, etc. 597pp. 5⅜ × 8½. 22632-8 Pa. $10.95

SUNDIALS, Their Theory and Construction, Albert Waugh. Far and away the best, most thorough coverage of ideas, mathematics concerned, types, construction, adjusting anywhere. Simple, nontechnical treatment allows even children to build several of these dials. Over 100 illustrations. 230pp. 5⅜ × 8½. 22947-5 Pa. $7.95

DYNAMICS OF FLUIDS IN POROUS MEDIA, Jacob Bear. For advanced students of ground water hydrology, soil mechanics and physics, drainage and irrigation engineering, and more. 335 illustrations. Exercises, with answers. 784pp. 6⅛ × 9¼. 65675-6 Pa. $19.95

SONGS OF EXPERIENCE: Facsimile Reproduction with 26 Plates in Full Color, William Blake. 26 full-color plates from a rare 1826 edition. Includes "The Tyger," "London," "Holy Thursday," and other poems. Printed text of poems. 48pp. 5¼ × 7. 24636-1 Pa. $4.95

OLD-TIME VIGNETTES IN FULL COLOR, Carol Belanger Grafton (ed.). Over 390 charming, often sentimental illustrations, selected from archives of Victorian graphics—pretty women posing, children playing, food, flowers, kittens and puppies, smiling cherubs, birds and butterflies, much more. All copyright-free. 48pp. 9¼ × 12¼. 27269-9 Pa. $5.95

PERSPECTIVE FOR ARTISTS, Rex Vicat Cole. Depth, perspective of sky and sea, shadows, much more, not usually covered. 391 diagrams, 81 reproductions of drawings and paintings. 279pp. 5⅜ × 8½. 22487-2 Pa. $6.95

DRAWING THE LIVING FIGURE, Joseph Sheppard. Innovative approach to artistic anatomy focuses on specifics of surface anatomy, rather than muscles and bones. Over 170 drawings of live models in front, back and side views, and in widely varying poses. Accompanying diagrams. 177 illustrations. Introduction. Index. 144pp. 8⅜ × 11¼. 26723-7 Pa. $8.95

GOTHIC AND OLD ENGLISH ALPHABETS: 100 Complete Fonts, Dan X. Solo. Add power, elegance to posters, signs, other graphics with 100 stunning copyright-free alphabets: Blackstone, Dolbey, Germania, 97 more—including many lower-case, numerals, punctuation marks. 104pp. 8⅛ × 11. 24695-7 Pa. $8.95

HOW TO DO BEADWORK, Mary White. Fundamental book on craft from simple projects to five-bead chains and woven works. 106 illustrations. 142pp. 5⅜ × 8. 20697-1 Pa. $4.95

THE BOOK OF WOOD CARVING, Charles Marshall Sayers. Finest book for beginners discusses fundamentals and offers 34 designs. "Absolutely first rate . . . well thought out and well executed."—E. J. Tangerman. 118pp. 7¾ × 10⅝. 23654-4 Pa. $5.95

ILLUSTRATED CATALOG OF CIVIL WAR MILITARY GOODS: Union Army Weapons, Insignia, Uniform Accessories, and Other Equipment, Schuyler, Hartley, and Graham. Rare, profusely illustrated 1846 catalog includes Union Army uniform and dress regulations, arms and ammunition, coats, insignia, flags, swords, rifles, etc. 226 illustrations. 160pp. 9 × 12. 24939-5 Pa. $10.95

WOMEN'S FASHIONS OF THE EARLY 1900s: An Unabridged Republication of "New York Fashions, 1909," National Cloak & Suit Co. Rare catalog of mail-order fashions documents women's and children's clothing styles shortly after the turn of the century. Captions offer full descriptions, prices. Invaluable resource for fashion, costume historians. Approximately 725 illustrations. 128pp. 8⅜ × 11¼. 27276-1 Pa. $11.95

THE 1912 AND 1915 GUSTAV STICKLEY FURNITURE CATALOGS, Gustav Stickley. With over 200 detailed illustrations and descriptions, these two catalogs are essential reading and reference materials and identification guides for Stickley furniture. Captions cite materials, dimensions and prices. 112pp. 6½ × 9¼. 26676-1 Pa. $9.95

EARLY AMERICAN LOCOMOTIVES, John H. White, Jr. Finest locomotive engravings from early 19th century: historical (1804–74), main-line (after 1870), special, foreign, etc. 147 plates. 142pp. 11⅜ × 8¼. 22772-3 Pa. $10.95

THE TALL SHIPS OF TODAY IN PHOTOGRAPHS, Frank O. Braynard. Lavishly illustrated tribute to nearly 100 majestic contemporary sailing vessels: Amerigo Vespucci, Clearwater, Constitution, Eagle, Mayflower, Sea Cloud, Victory, many more. Authoritative captions provide statistics, background on each ship. 190 black-and-white photographs and illustrations. Introduction. 128pp. 8⅞ × 11¾. 27163-3 Pa. $13.95

EARLY NINETEENTH-CENTURY CRAFTS AND TRADES, Peter Stockham (ed.). Extremely rare 1807 volume describes to youngsters the crafts and trades of the day: brickmaker, weaver, dressmaker, bookbinder, ropemaker, saddler, many more. Quaint prose, charming illustrations for each craft. 20 black-and-white line illustrations. 192pp. 4⅝ × 6.   27293-1 Pa. $4.95

VICTORIAN FASHIONS AND COSTUMES FROM HARPER'S BAZAR, 1867–1898, Stella Blum (ed.). Day costumes, evening wear, sports clothes, shoes, hats, other accessories in over 1,000 detailed engravings. 320pp. 9⅜ × 12¼.
22990-4 Pa. $13.95

GUSTAV STICKLEY, THE CRAFTSMAN, Mary Ann Smith. Superb study surveys broad scope of Stickley's achievement, especially in architecture. Design philosophy, rise and fall of the Craftsman empire, descriptions and floor plans for many Craftsman houses, more. 86 black-and-white halftones. 31 line illustrations. Introduction. 208pp. 6½ × 9¼.   27210-9 Pa. $9.95

THE LONG ISLAND RAIL ROAD IN EARLY PHOTOGRAPHS, Ron Ziel. Over 220 rare photos, informative text document origin (1844) and development of rail service on Long Island. Vintage views of early trains, locomotives, stations, passengers, crews, much more. Captions. 8⅞ × 11¾.   26301-0 Pa. $13.95

THE BOOK OF OLD SHIPS: From Egyptian Galleys to Clipper Ships, Henry B. Culver. Superb, authoritative history of sailing vessels, with 80 magnificent line illustrations. Galley, bark, caravel, longship, whaler, many more. Detailed, informative text on each vessel by noted naval historian. Introduction. 256pp. 5⅝ × 8½.   27332-6 Pa. $6.95

TEN BOOKS ON ARCHITECTURE, Vitruvius. The most important book ever written on architecture. Early Roman aesthetics, technology, classical orders, site selection, all other aspects. Morgan translation. 331pp. 5⅜ × 8½. 20645-9 Pa. $8.95

THE HUMAN FIGURE IN MOTION, Eadweard Muybridge. More than 4,500 stopped-action photos, in action series, showing undraped men, women, children jumping, lying down, throwing, sitting, wrestling, carrying, etc. 390pp. 7⅞ × 10⅝.
20204-6 Clothbd. $24.95

TREES OF THE EASTERN AND CENTRAL UNITED STATES AND CANADA, William M. Harlow. Best one-volume guide to 140 trees. Full descriptions, woodlore, range, etc. Over 600 illustrations. Handy size. 288pp. 4½ × 6⅜.
20395-6 Pa. $5.95

SONGS OF WESTERN BIRDS, Dr. Donald J. Borror. Complete song and call repertoire of 60 western species, including flycatchers, juncoes, cactus wrens, many more—includes fully illustrated booklet.   Cassette and manual 99913-0 $8.95

GROWING AND USING HERBS AND SPICES, Milo Miloradovich. Versatile handbook provides all the information needed for cultivation and use of all the herbs and spices available in North America. 4 illustrations. Index. Glossary. 236pp. 5⅜ × 8½.   25058-X Pa. $6.95

BIG BOOK OF MAZES AND LABYRINTHS, Walter Shepherd. 50 mazes and labyrinths in all—classical, solid, ripple, and more—in one great volume. Perfect inexpensive puzzler for clever youngsters. Full solutions. 112pp. 8⅛ × 11.
22951-3 Pa. $4.95

PIANO TUNING, J. Cree Fischer. Clearest, best book for beginner, amateur. Simple repairs, raising dropped notes, tuning by easy method of flattened fifths. No previous skills needed. 4 illustrations. 201pp. 5⅜ × 8½.　23267-0 Pa. $5.95

A SOURCE BOOK IN THEATRICAL HISTORY, A. M. Nagler. Contemporary observers on acting, directing, make-up, costuming, stage props, machinery, scene design, from Ancient Greece to Chekhov. 611pp. 5⅜ × 8½.　20515-0 Pa. $11.95

THE COMPLETE NONSENSE OF EDWARD LEAR, Edward Lear. All nonsense limericks, zany alphabets, Owl and Pussycat, songs, nonsense botany, etc., illustrated by Lear. Total of 320pp. 5⅜ × 8½. (USO)　20167-8 Pa. $6.95

VICTORIAN PARLOUR POETRY: An Annotated Anthology, Michael R. Turner. 117 gems by Longfellow, Tennyson, Browning, many lesser-known poets. "The Village Blacksmith," "Curfew Must Not Ring Tonight," "Only a Baby Small," dozens more, often difficult to find elsewhere. Index of poets, titles, first lines. xxiii + 325pp. 5⅜ × 8¼.　27044-0 Pa. $8.95

DUBLINERS, James Joyce. Fifteen stories offer vivid, tightly focused observations of the lives of Dublin's poorer classes. At least one, "The Dead," is considered a masterpiece. Reprinted complete and unabridged from standard edition. 160pp. 5³/₁₆ × 8¼.　26870-5 Pa. $1.00

THE HAUNTED MONASTERY and THE CHINESE MAZE MURDERS, Robert van Gulik. Two full novels by van Gulik, set in 7th-century China, continue adventures of Judge Dee and his companions. An evil Taoist monastery, seemingly supernatural events; overgrown topiary maze hides strange crimes. 27 illustrations. 328pp. 5⅜ × 8½.　23502-5 Pa. $7.95

THE BOOK OF THE SACRED MAGIC OF ABRAMELIN THE MAGE, translated by S. MacGregor Mathers. Medieval manuscript of ceremonial magic. Basic document in Aleister Crowley, Golden Dawn groups. 268pp. 5⅜ × 8½.
23211-5 Pa. $8.95

NEW RUSSIAN-ENGLISH AND ENGLISH-RUSSIAN DICTIONARY, M. A. O'Brien. This is a remarkably handy Russian dictionary, containing a surprising amount of information, including over 70,000 entries. 366pp. 4½ × 6⅛.
20208-9 Pa. $9.95

HISTORIC HOMES OF THE AMERICAN PRESIDENTS, Second, Revised Edition, Irvin Haas. A traveler's guide to American Presidential homes, most open to the public, depicting and describing homes occupied by every American President from George Washington to George Bush. With visiting hours, admission charges, travel routes. 175 photographs. Index. 160pp. 8¼ × 11. 26751-2 Pa. $10.95

NEW YORK IN THE FORTIES, Andreas Feininger. 162 brilliant photographs by the well-known photographer, formerly with *Life* magazine. Commuters, shoppers, Times Square at night, much else from city at its peak. Captions by John von Hartz. 181pp. 9¼ × 10¾.　23585-8 Pa. $12.95

INDIAN SIGN LANGUAGE, William Tomkins. Over 525 signs developed by Sioux and other tribes. Written instructions and diagrams. Also 290 pictographs. 111pp. 6⅛ × 9¼.　22029-X Pa. $3.50

ANATOMY: A Complete Guide for Artists, Joseph Sheppard. A master of figure drawing shows artists how to render human anatomy convincingly. Over 460 illustrations. 224pp. 8⅜ × 11¼. 27279-6 Pa. $10.95

MEDIEVAL CALLIGRAPHY: Its History and Technique, Marc Drogin. Spirited history, comprehensive instruction manual covers 13 styles (ca. 4th century thru 15th). Excellent photographs; directions for duplicating medieval techniques with modern tools. 224pp. 8⅜ × 11¼. 26142-5 Pa. $11.95

DRIED FLOWERS: How to Prepare Them, Sarah Whitlock and Martha Rankin. Complete instructions on how to use silica gel, meal and borax, perlite aggregate, sand and borax, glycerine and water to create attractive permanent flower arrangements. 12 illustrations. 32pp. 5⅜ × 8½. 21802-3 Pa. $1.00

EASY-TO-MAKE BIRD FEEDERS FOR WOODWORKERS, Scott D. Campbell. Detailed, simple-to-use guide for designing, constructing, caring for and using feeders. Text, illustrations for 12 classic and contemporary designs. 96pp. 5⅜ × 8½. 25847-5 Pa. $2.95

OLD-TIME CRAFTS AND TRADES, Peter Stockham. An 1807 book created to teach children about crafts and trades open to them as future careers. It describes in detailed, nontechnical terms 24 different occupations, among them coachmaker, gardener, hairdresser, lacemaker, shoemaker, wheelwright, copper-plate printer, milliner, trunkmaker, merchant and brewer. Finely detailed engravings illustrate each occupation. 192pp. 4⅝ × 6. 27398-9 Pa. $4.95

THE HISTORY OF UNDERCLOTHES, C. Willett Cunnington and Phyllis Cunnington. Fascinating, well-documented survey covering six centuries of English undergarments, enhanced with over 100 illustrations: 12th-century laced-up bodice, footed long drawers (1795), 19th-century bustles, 19th-century corsets for men, Victorian "bust improvers," much more. 272pp. 5⅜ × 8¼. 27124-2 Pa. $9.95

ARTS AND CRAFTS FURNITURE: The Complete Brooks Catalog of 1912, Brooks Manufacturing Co. Photos and detailed descriptions of more than 150 now very collectible furniture designs from the Arts and Crafts movement depict davenports, settees, buffets, desks, tables, chairs, bedsteads, dressers and more, all built of solid, quarter-sawed oak. Invaluable for students and enthusiasts of antiques, Americana and the decorative arts. 80pp. 6½ × 9¼. 27471-3 Pa. $7.95

HOW WE INVENTED THE AIRPLANE: An Illustrated History, Orville Wright. Fascinating firsthand account covers early experiments, construction of planes and motors, first flights, much more. Introduction and commentary by Fred C. Kelly. 76 photographs. 96pp. 8¼ × 11. 25662-6 Pa. $8.95

THE ARTS OF THE SAILOR: Knotting, Splicing and Ropework, Hervey Garrett Smith. Indispensable shipboard reference covers tools, basic knots and useful hitches; handsewing and canvas work, more. Over 100 illustrations. Delightful reading for sea lovers. 256pp. 5⅜ × 8½. 26440-8 Pa. $7.95

FRANK LLOYD WRIGHT'S FALLINGWATER: The House and Its History, Second, Revised Edition, Donald Hoffmann. A total revision—both in text and illustrations—of the standard document on Fallingwater, the boldest, most personal architectural statement of Wright's mature years, updated with valuable new material from the recently opened Frank Lloyd Wright Archives. "Fascinating"—*The New York Times.* 116 illustrations. 128pp. 9¼ × 10¾.
27430-6 Pa. $10.95

PHOTOGRAPHIC SKETCHBOOK OF THE CIVIL WAR, Alexander Gardner. 100 photos taken on field during the Civil War. Famous shots of Manassas, Harper's Ferry, Lincoln, Richmond, slave pens, etc. 244pp. 10⅝ × 8¼.
22731-6 Pa. $9.95

FIVE ACRES AND INDEPENDENCE, Maurice G. Kains. Great back-to-the-land classic explains basics of self-sufficient farming. The one book to get. 95 illustrations. 397pp. 5⅜ × 8½. 20974-1 Pa. $7.95

SONGS OF EASTERN BIRDS, Dr. Donald J. Borror. Songs and calls of 60 species most common to eastern U.S.: warblers, woodpeckers, flycatchers, thrushes, larks, many more in high-quality recording. Cassette and manual 99912-2 $8.95

A MODERN HERBAL, Margaret Grieve. Much the fullest, most exact, most useful compilation of herbal material. Gigantic alphabetical encyclopedia, from aconite to zedoary, gives botanical information, medical properties, folklore, economic uses, much else. Indispensable to serious reader. 161 illustrations. 888pp. 6½ × 9¼. 2-vol. set. (USO) Vol. I: 22798-7 Pa. $9.95
Vol. II: 22799-5 Pa. $9.95

HIDDEN TREASURE MAZE BOOK, Dave Phillips. Solve 34 challenging mazes accompanied by heroic tales of adventure. Evil dragons, people-eating plants, bloodthirsty giants, many more dangerous adversaries lurk at every twist and turn. 34 mazes, stories, solutions. 48pp. 8¼ × 11. 24566-7 Pa. $2.95

LETTERS OF W. A. MOZART, Wolfgang A. Mozart. Remarkable letters show bawdy wit, humor, imagination, musical insights, contemporary musical world; includes some letters from Leopold Mozart. 276pp. 5⅜ × 8½. 22859-2 Pa. $7.95

BASIC PRINCIPLES OF CLASSICAL BALLET, Agrippina Vaganova. Great Russian theoretician, teacher explains methods for teaching classical ballet. 118 illustrations. 175pp. 5⅜ × 8½. 22036-2 Pa. $4.95

THE JUMPING FROG, Mark Twain. Revenge edition. The original story of The Celebrated Jumping Frog of Calaveras County, a hapless French translation, and Twain's hilarious "retranslation" from the French. 12 illustrations. 66pp. 5⅜ × 8½.
22686-7 Pa. $3.95

BEST REMEMBERED POEMS, Martin Gardner (ed.). The 126 poems in this superb collection of 19th- and 20th-century British and American verse range from Shelley's "To a Skylark" to the impassioned "Renascence" of Edna St. Vincent Millay and to Edward Lear's whimsical "The Owl and the Pussycat." 224pp. 5⅜ × 8½.
27165-X Pa. $4.95

COMPLETE SONNETS, William Shakespeare. Over 150 exquisite poems deal with love, friendship, the tyranny of time, beauty's evanescence, death and other themes in language of remarkable power, precision and beauty. Glossary of archaic terms. 80pp. 5³⁄₁₆ × 8¼. 26686-9 Pa. $1.00

BODIES IN A BOOKSHOP, R. T. Campbell. Challenging mystery of blackmail and murder with ingenious plot and superbly drawn characters. In the best tradition of British suspense fiction. 192pp. 5⅜ × 8½. 24720-1 Pa. $5.95

THE WIT AND HUMOR OF OSCAR WILDE, Alvin Redman (ed.). More than 1,000 ripostes, paradoxes, wisecracks: Work is the curse of the drinking classes; I can resist everything except temptation; etc. 258pp. 5⅜ × 8½. 20602-5 Pa. $5.95

SHAKESPEARE LEXICON AND QUOTATION DICTIONARY, Alexander Schmidt. Full definitions, locations, shades of meaning in every word in plays and poems. More than 50,000 exact quotations. 1,485pp. 6½ × 9¼. 2-vol. set.
Vol. I: 22726-X Pa. $16.95
Vol. 2: 22727-8 Pa. $15.95

SELECTED POEMS, Emily Dickinson. Over 100 best-known, best-loved poems by one of America's foremost poets, reprinted from authoritative early editions. No comparable edition at this price. Index of first lines. 64pp. 5³⁄₁₆ × 8¼.
26466-1 Pa. $1.00

CELEBRATED CASES OF JUDGE DEE (DEE GOONG AN), translated by Robert van Gulik. Authentic 18th-century Chinese detective novel; Dee and associates solve three interlocked cases. Led to van Gulik's own stories with same characters. Extensive introduction. 9 illustrations. 237pp. 5⅜ × 8½.
23337-5 Pa. $6.95

THE MALLEUS MALEFICARUM OF KRAMER AND SPRENGER, translated by Montague Summers. Full text of most important witchhunter's "bible," used by both Catholics and Protestants. 278pp. 6⅝ × 10. 22802-9 Pa. $11.95

SPANISH STORIES/CUENTOS ESPAÑOLES: A Dual-Language Book, Angel Flores (ed.). Unique format offers 13 great stories in Spanish by Cervantes, Borges, others. Faithful English translations on facing pages. 352pp. 5⅜ × 8½.
25399-6 Pa. $8.95

THE CHICAGO WORLD'S FAIR OF 1893: A Photographic Record, Stanley Appelbaum (ed.). 128 rare photos show 200 buildings, Beaux-Arts architecture, Midway, original Ferris Wheel, Edison's kinetoscope, more. Architectural emphasis; full text. 116pp. 8¼ × 11. 23990-X Pa. $9.95

OLD QUEENS, N.Y., IN EARLY PHOTOGRAPHS, Vincent F. Seyfried and William Asadorian. Over 160 rare photographs of Maspeth, Jamaica, Jackson Heights, and other areas. Vintage views of DeWitt Clinton mansion, 1939 World's Fair and more. Captions. 192pp. 8⅞ × 11. 26358-4 Pa. $12.95

CAPTURED BY THE INDIANS: 15 Firsthand Accounts, 1750–1870, Frederick Drimmer. Astounding true historical accounts of grisly torture, bloody conflicts, relentless pursuits, miraculous escapes and more, by people who lived to tell the tale. 384pp. 5⅜ × 8½. 24901-8 Pa. $8.95

THE WORLD'S GREAT SPEECHES, Lewis Copeland and Lawrence W. Lamm (eds.). Vast collection of 278 speeches of Greeks to 1970. Powerful and effective models; unique look at history. 842pp. 5⅜ × 8½. 20468-5 Pa. $14.95

THE BOOK OF THE SWORD, Sir Richard F. Burton. Great Victorian scholar/adventurer's eloquent, erudite history of the "queen of weapons"—from prehistory to early Roman Empire. Evolution and development of early swords, variations (sabre, broadsword, cutlass, scimitar, etc.), much more. 336pp. 6⅛ × 9¼. 25434-8 Pa. $8.95

AUTOBIOGRAPHY: The Story of My Experiments with Truth, Mohandas K. Gandhi. Boyhood, legal studies, purification, the growth of the Satyagraha (nonviolent protest) movement. Critical, inspiring work of the man responsible for the freedom of India. 480pp. 5⅜ × 8½. (USO) **24593-4 Pa. $8.95**

CELTIC MYTHS AND LEGENDS, T. W. Rolleston. Masterful retelling of Irish and Welsh stories and tales. Cuchulain, King Arthur, Deirdre, the Grail, many more. First paperback edition. 58 full-page illustrations. 512pp. 5⅜ × 8½. 26507-2 Pa. $9.95

THE PRINCIPLES OF PSYCHOLOGY, William James. Famous long course complete, unabridged. Stream of thought, time perception, memory, experimental methods; great work decades ahead of its time. 94 figures. 1,391pp. 5⅜ × 8½. 2-vol. set.
Vol. I: 20381-6 Pa. $12.95
Vol. II: 20382-4 Pa. $12.95

THE WORLD AS WILL AND REPRESENTATION, Arthur Schopenhauer. Definitive English translation of Schopenhauer's life work, correcting more than 1,000 errors, omissions in earlier translations. Translated by E. F. J. Payne. Total of 1,269pp. 5⅜ × 8½. 2-vol. set. Vol. 1: 21761-2 Pa. $11.95
Vol. 2: 21762-0 Pa. $11.95

MAGIC AND MYSTERY IN TIBET, Madame Alexandra David-Neel. Experiences among lamas, magicians, sages, sorcerers, Bonpa wizards. A true psychic discovery. 32 illustrations. 321pp. 5⅜ × 8½. (USO) 22682-4 Pa. $8.95

THE EGYPTIAN BOOK OF THE DEAD, E. A. Wallis Budge. Complete reproduction of Ani's papyrus, finest ever found. Full hieroglyphic text, interlinear transliteration, word-for-word translation, smooth translation. 533pp. 6½ × 9¼. 21866-X Pa. $9.95

MATHEMATICS FOR THE NONMATHEMATICIAN, Morris Kline. Detailed, college-level treatment of mathematics in cultural and historical context, with numerous exercises. Recommended Reading Lists. Tables. Numerous figures. 641pp. 5⅜ × 8½. 24823-2 Pa. $11.95

THEORY OF WING SECTIONS: Including a Summary of Airfoil Data, Ira H. Abbott and A. E. von Doenhoff. Concise compilation of subsonic aerodynamic characteristics of NACA wing sections, plus description of theory. 350pp. of tables. 693pp. 5⅜ × 8½. 60586-8 Pa. $14.95

THE RIME OF THE ANCIENT MARINER, Gustave Doré, S. T. Coleridge. Doré's finest work; 34 plates capture moods, subtleties of poem. Flawless full-size reproductions printed on facing pages with authoritative text of poem. "Beautiful. Simply beautiful."—*Publisher's Weekly*. 77pp. 9¼ × 12. 22305-1 Pa. $6.95

NORTH AMERICAN INDIAN DESIGNS FOR ARTISTS AND CRAFTSPEOPLE, Eva Wilson. Over 360 authentic copyright-free designs adapted from Navajo blankets, Hopi pottery, Sioux buffalo hides, more. Geometrics, symbolic figures, plant and animal motifs, etc. 128pp. 8⅜ × 11. (EUK) 25341-4 Pa. $7.95

SCULPTURE: Principles and Practice, Louis Slobodkin. Step-by-step approach to clay, plaster, metals, stone; classical and modern. 253 drawings, photos. 255pp. 8⅛ × 11. 22960-2 Pa. $10.95

THE INFLUENCE OF SEA POWER UPON HISTORY, 1660-1783, A. T. Mahan. Influential classic of naval history and tactics still used as text in war colleges. First paperback edition. 4 maps. 24 battle plans. 640pp. 5⅜ × 8½.
25509-3 Pa. $12.95

THE STORY OF THE TITANIC AS TOLD BY ITS SURVIVORS, Jack Winocour (ed.). What it was really like. Panic, despair, shocking inefficiency, and a little heroism. More thrilling than any fictional account. 26 illustrations. 320pp. 5⅜ × 8½. 20610-6 Pa. $8.95

FAIRY AND FOLK TALES OF THE IRISH PEASANTRY, William Butler Yeats (ed.). Treasury of 64 tales from the twilight world of Celtic myth and legend: "The Soul Cages," "The Kildare Pooka," "King O'Toole and his Goose," many more. Introduction and Notes by W. B. Yeats. 352pp. 5⅜ × 8½. 26941-8 Pa. $8.95

BUDDHIST MAHAYANA TEXTS, E. B. Cowell and Others (eds.). Superb, accurate translations of basic documents in Mahayana Buddhism, highly important in history of religions. The Buddha-karita of Asvaghosha, Larger Sukhavativyuha, more. 448pp. 5⅜ × 8½. , 25552-2 Pa. $9.95

ONE TWO THREE . . . INFINITY: Facts and Speculations of Science, George Gamow. Great physicist's fascinating, readable overview of contemporary science: number theory, relativity, fourth dimension, entropy, genes, atomic structure, much more. 128 illustrations. Index. 352pp. 5⅜ × 8½. 25664-2 Pa. $8.95

ENGINEERING IN HISTORY, Richard Shelton Kirby, et al. Broad, nontechnical survey of history's major technological advances: birth of Greek science, industrial revolution, electricity and applied science, 20th-century automation, much more. 181 illustrations. ". . . excellent . . ."—Isis. Bibliography. vii + 530pp. 5⅜ × 8¼.
26412-2 Pa. $14.95

Prices subject to change without notice.

Available at your book dealer or write for free catalog to Dept. GI, Dover Publications, Inc., 31 East 2nd St., Mineola, N.Y. 11501. Dover publishes more than 500 books each year on science, elementary and advanced mathematics, biology, music, art, literary history, social sciences and other areas.